SpringerBriefs in Applied Sciences and Technology

SpringerBriefs present concise summaries of cutting-edge research and practical applications across a wide spectrum of fields. Featuring compact volumes of 50 to 125 pages, the series covers a range of content from professional to academic.

Typical publications can be:

- A timely report of state-of-the art methods
- An introduction to or a manual for the application of mathematical or computer techniques
- A bridge between new research results, as published in journal articles
- A snapshot of a hot or emerging topic
- An in-depth case study
- A presentation of core concepts that students must understand in order to make independent contributions

SpringerBriefs are characterized by fast, global electronic dissemination, standard publishing contracts, standardized manuscript preparation and formatting guidelines, and expedited production schedules.

On the one hand, **SpringerBriefs in Applied Sciences and Technology** are devoted to the publication of fundamentals and applications within the different classical engineering disciplines as well as in interdisciplinary fields that recently emerged between these areas. On the other hand, as the boundary separating fundamental research and applied technology is more and more dissolving, this series is particularly open to trans-disciplinary topics between fundamental science and engineering.

Indexed by EI-Compendex, SCOPUS and Springerlink.

Prasastha Vemula · Ramesh Chatragadda

Ecofriendly and Multifunctional Nanoparticles

Synthesis from Seaweeds, Seagrasses, Mangroves, and Other Coastal Plants

Prasastha Vemula
Animal and Fisheries Science Section
ICAR—Central Coastal Agricultural
Research Institute
Velha, Goa, India

Ramesh Chatragadda
Academy of Scientific and Innovative
Research (AcSIR)
Ghaziabad, Uttar Pradesh, India

Biological Oceanography Division (BOD)
CSIR-National Institute of Oceanography
(CSIR-NIO)
Panaji, Goa, India

ISSN 2191-530X   ISSN 2191-5318 (electronic)
SpringerBriefs in Applied Sciences and Technology
ISBN 978-981-96-5877-0   ISBN 978-981-96-5878-7 (eBook)
https://doi.org/10.1007/978-981-96-5878-7

© The Editor(s) (if applicable) and The Author(s), under exclusive license to Springer Nature Singapore Pte Ltd. 2025

This work is subject to copyright. All rights are solely and exclusively licensed by the Publisher, whether the whole or part of the material is concerned, specifically the rights of translation, reprinting, reuse of illustrations, recitation, broadcasting, reproduction on microfilms or in any other physical way, and transmission or information storage and retrieval, electronic adaptation, computer software, or by similar or dissimilar methodology now known or hereafter developed.

The use of general descriptive names, registered names, trademarks, service marks, etc. in this publication does not imply, even in the absence of a specific statement, that such names are exempt from the relevant protective laws and regulations and therefore free for general use.

The publisher, the authors and the editors are safe to assume that the advice and information in this book are believed to be true and accurate at the date of publication. Neither the publisher nor the authors or the editors give a warranty, expressed or implied, with respect to the material contained herein or for any errors or omissions that may have been made. The publisher remains neutral with regard to jurisdictional claims in published maps and institutional affiliations.

This Springer imprint is published by the registered company Springer Nature Singapore Pte Ltd.
The registered company address is: 152 Beach Road, #21-01/04 Gateway East, Singapore 189721, Singapore

If disposing of this product, please recycle the paper.

We are indeed very much excited to dedicate our first book to our parents, whose unwavering encouragement and never-ending support have been the cornerstone of our academics and research pursuits. Their unending love and selfless contributions are a debt we can never fully repay.

Preface

Nanoparticles have demonstrated multifaceted application in numerous fields. Their source of synthesis reported to play major role in determining effective functionality in specified applications. Though several studies have reported nanoparticles synthesis from various plants and other sources, marine flora derived nanoparticles (MFNPs) and their applications have not been detailed in any book. Thus, we address this gap and provide a comprehensive information on MFNPs for the benefit of researchers, students, industrial personnel, and medical professionals. While this first edition would serve as a database and base line book of MFNPs, we further aim to develop a database of MFNPs by adding A to Z information available across the world.

Velha, India	Prasastha Vemula
Ghaziabad/Panaji, India	Ramesh Chatragadda

Acknowledgements Prasastha Vemula thank the Director ICAR-CCARI for his constant encouragement and support in the research field. Ramesh Chatragadda is grateful to the Director CSIR-NIO for his support and motivation. This is CSIR-NIO's reference number: 11358. Some icons used in figures were obtained from free online icon sources: https://www.phylopic.org/ and https://www.iconfinder.com/, for illustration purposes.

Competing Interests The authors have no competing interests to declare that are relevant to the content of this manuscript.

About This Book

This book highlights a comprehensive analysis of marine flora-derived nanoparticles (MFNPs), encompassing synthesis, characterization, physiological properties, and regulatory factors. It delves into the benefits of conjugation, mechanisms of action, and limitations, highlighting the challenges and opportunities in the field. The book identifies research gaps and outlines future directions, providing valuable insights for researchers and professionals working with MFNPs. The book covers a range of marine flora, including seaweeds, seagrasses, mangroves, and other coastal plants. By shedding light on the potential applications of MFNPs, this book serves as a valuable resource for those in the field. It fosters further research and innovation, ultimately contributing to the advancement of MFNP-based technologies and their applications in various industries.

Contents

1 **Ecofriendly Nanoparticles from Marine Flora: Green Synthesis, Applications, and Future Prospects** 1
   1.1 Introduction 2
       1.1.1 Emphasis of Marine Plants in Ancient Literature 3
       1.1.2 Benefits of Marine Flora-Derived Nanoparticles Over Terrestrial Plants 3
   1.2 Literature Source and Search Strategy 4
   1.3 Advantages of Nanoparticles Derived from Marine Flora 4
   1.4 Reducing and Capping Agents in Marine Flora 5
   1.5 Green Synthesis of MFNPs 5
   1.6 Morphology and Identification of Nanoparticles 44
   1.7 Factors Affecting the Synthesis of Nanoparticles 47
   1.8 Mechanism of Nanoparticle Synthesis from Marine Flora 48
   1.9 Applications of Marine Flora-Derived Nanoparticles (MFNPs) 50
       1.9.1 Antiviral Activity 50
       1.9.2 Antibacterial Properties 50
       1.9.3 Antituberculosis Activity 84
       1.9.4 Antifungal Properties 84
       1.9.5 Anticancer and Cytotoxic Properties 86
       1.9.6 Antidiabetic Activity 102
       1.9.7 Wound Healing Properties 102
       1.9.8 Anti-inflammatory Properties 103
       1.9.9 Anti-Alzheimer Activity 103
       1.9.10 Antiprotozoal/Antiplasmodial Activity 103
       1.9.11 Antifouling Activity Nanoparticles 103
       1.9.12 Mosquitocidal Activity 104
       1.9.13 Insecticidal Activity 105
       1.9.14 Seed Germination Properties 106
       1.9.15 Biopreservation Properties 106
       1.9.16 Applications in Textile Industry 107

	1.9.17	Biosensor and Bioremediation Properties	108
	1.9.18	Fluorescence Enhancement or Quenching Application	108
	1.9.19	Electrocatalyst Properties	109
	1.9.20	Other Biological Properties	109
1.10	Enhancing Biological Properties of Nanoparticles by Conjugation		110
1.11	Limitations of Using Marine Plants for Synthesis of Nanoparticles		110
1.12	Research Gaps		111
1.13	Conclusion		111
References			112

About the Authors

Dr. Prasastha Vemula is a veterinary doctor who earned her Ph.D. from ICAR-Indian Veterinary Research Institute (IVRI), India. She has conducted extensive research on phytochemicals and phyto nanodrugs to combat Antimicrobial Resistance (AMR) displaying clinical bacteria. She has several international research publications in this field demonstrating the need of exploration of green nanodrugs for various applications. Currently, she is working at ICAR-Central Coastal Agricultural Research Institute, Goa, India. Her curiosity to explore the marine plants derived nanoparticles has been a significant milestone in her research journey, offering valuable insights that are detailed in this book.

Dr. Ramesh Chatragadda is a marine biologist and enthusiastic explorer of marine life, dedicated to uncovering its potential for various industrial and societal benefits. He obtained Ph.D. from Pondicherry University, Andaman Campus and currently working as a senior scientist at CSIR-national Institute of Oceanography, Goa. Since 2011, he has been exploring the diversity, distribution and bioactive compounds of marine microbes, plants, and animals from both coastal and deep-sea environments, focusing on their applications in combating multidrug-resistant microbes. His research journey has resulted in over 100 publications. This book is one of the major accomplishments in his research, revealing the importance of marine plant-derived nanoparticles in improving human health.

Abbreviations

AChE	Acetylcholinesterase
AgClNPs	Silver chloride nanoparticles
AgNPs	Silver nanoparticles
Ag-ZnONPs	Silver-zinc oxide nanoparticles
AuNPs	Gold nanoparticles
CC_{50}	Cytotoxic concentration 50%
CdO	Cadmium oxide
CdO-ZnONPs	Cadmium oxide and zinc oxide nanoparticles
CdSNPs	Cadmium sulfide nanoparticles
CoONPs/Co_3O_4NPs	Cobalt oxide nanoparticles
CuNPs	Copper nanoparticles
CuONPs	Copper oxide nanoparticles
DEN-2	Dengue serotype 2
DLS	Dynamic light scattering
DNA	Deoxyribonucleic acid
DOX	Doxorubicin
DOX-AcFuNPs	Doxorubicin-loaded acetylated fucoidan nanoparticles
DSP	Deoiled *Saccharina japonica* powder
EC_{50}	Half maximal effective concentration
EDX	Energy-dispersive X-ray
EPI-CAO-AuNPs	Epirubicin-loaded kappa-carrageenan gold nanoparticles
FCD/LFNPs	Fucoidan/Lactoferrin nanoparticles
FCD/QCNPs	Fucoidan/Quinacrine nanoparticles
FeNPs	Iron nanoparticles
FeONPs/Fe_3O_4NPs	Iron oxide nanoparticles
FTIR	Fourier Transform Infrared Spectroscopy
Fu/CHNPs	Fucoidan/chitosan nanoparticles
GI_{50}	Growth inhibition 50%
HIV	Human Immunodeficiency Virus
IC_{50}	Half-maximal inhibitory concentration
LC_{50}	Lethal Concentration 50%

LC_{90}	Lethal Concentration 90%
MFNPs	Marine flora-derived nanoparticles
MgONPs	Magnesium oxide nanoparticles
MIC	Minimum inhibitory concentration
mL	Milliliter
MLC	Minimum lethal concentration
mM	Millimolar
NPs	Nanoparticles
PdNPs	Palladium nanoparticles
PEI-FCD-DOX NPs	Polyethylenimine-fucoidan-Doxorubicin nanoparticles
pH	Potential of hydrogen
PtNPs	Platinum nanoparticles
ROS	Reactive oxygen species
SAED	Selected area electron diffraction
SEM	Scanning electron microscope
SeNPs	Selenium nanoparticles
SiO_2	Silicon dioxide nanoparticles
SPIONPs	Super-paramagnetic iron oxide nanoparticles
TEM	Transmission Electron Microscopy
TiO_2NPs	Titanium dioxide nanoparticles
ULANP	*Ulva lactuca* algae extract-loaded albumin nanoparticles
UV-Vis	Ultraviolet–visible
X-RD	X-ray Diffraction
ZnNPs	Zinc nanoparticles
ZnONPs	Zinc oxide nanoparticles
ZrO_2NP	Zirconium dioxide nanoparticles
ZrONPs	Zirconium oxide nanoparticles
μg	Microgram
μL	Microliter

List of Figures

Fig. 1.1	Some reducing and capping agents originated from of marine plants	6
Fig. 1.2	Illustration depicting methodology flow involved in green synthesis of nanoparticles from marine plants	42
Fig. 1.3	Illustration depicting a detailed flow chart of methodology involved in green synthesis of nanoparticles from marine plants. Icons are obtained from Iconfinder.com for illustration purpose	43
Fig. 1.4	Morphology of various nanoparticles derived from marine plant extracts	46
Fig. 1.5	Various factors that affect synthesis of nanoparticles from marine plant extracts. Icon credits: https://www.flaticon.com	49
Fig. 1.6	Mechanism of antimicrobial and anticancer activity demonstrated by NPs fabricated from marine plants. Illustration is created based on the following literature (Ajarem et al. 2022; Baskar et al. 2023; Choudhary et al. 2022; Fouda et al. 2022b; Puskulluoglu and Michalak, 2022). Few icons are taken from Iconfinder.com for illustration purpose	51
Fig. 1.7	Number of reports on nanoparticles synthesized from each marine floral group	82
Fig. 1.8	Illustration depicting the applications of various nanoparticles derived from green seaweeds	96
Fig. 1.9	Illustration depicting the applications of various nanoparticles derived from brown seaweeds	97
Fig. 1.10	Illustration depicting the applications of various nanoparticles derived from red seaweeds	98
Fig. 1.11	Illustration depicting the applications of various nanoparticles derived from mangroves	99

Fig. 1.12	Illustration depicting the applications of various nanoparticles derived from seagrasses and other coastal plants	100
Fig. 1.13	Illustration depicting the applications of various nanoparticles derived from other coastal plants	101

List of Tables

Table 1.1	Details of green synthesized nanoparticles from marine plants	7
Table 1.2	Antimicrobial properties of marine plants derived nanoparticles against different pathogenic microorganisms	52
Table 1.3	Anticancer and cytotoxic activity of marine plants derived nanoparticles against different cell lines	87

Chapter 1
Ecofriendly Nanoparticles from Marine Flora: Green Synthesis, Applications, and Future Prospects

Contents

1.1	Introduction	2
	1.1.1 Emphasis of Marine Plants in Ancient Literature	3
	1.1.2 Benefits of Marine Flora-Derived Nanoparticles Over Terrestrial Plants	3
1.2	Literature Source and Search Strategy	4
1.3	Advantages of Nanoparticles Derived from Marine Flora	4
1.4	Reducing and Capping Agents in Marine Flora	5
1.5	Green Synthesis of MFNPs	5
1.6	Morphology and Identification of Nanoparticles	44
1.7	Factors Affecting the Synthesis of Nanoparticles	47
1.8	Mechanism of Nanoparticle Synthesis from Marine Flora	48
1.9	Applications of Marine Flora-Derived Nanoparticles (MFNPs)	50
	1.9.1 Antiviral Activity	50
	1.9.2 Antibacterial Properties	50
	1.9.3 Antituberculosis Activity	84
	1.9.4 Antifungal Properties	84
	1.9.5 Anticancer and Cytotoxic Properties	86
	1.9.6 Antidiabetic Activity	102
	1.9.7 Wound Healing Properties	102
	1.9.8 Anti-inflammatory Properties	103
	1.9.9 Anti-Alzheimer Activity	103
	1.9.10 Antiprotozoal/Antiplasmodial Activity	103
	1.9.11 Antifouling Activity Nanoparticles	103
	1.9.12 Mosquitocidal Activity	104
	1.9.13 Insecticidal Activity	105
	1.9.14 Seed Germination Properties	106
	1.9.15 Biopreservation Properties	106
	1.9.16 Applications in Textile Industry	107
	1.9.17 Biosensor and Bioremediation Properties	108
	1.9.18 Fluorescence Enhancement or Quenching Application	108
	1.9.19 Electrocatalyst Properties	109
	1.9.20 Other Biological Properties	109
1.10	Enhancing Biological Properties of Nanoparticles by Conjugation	110
1.11	Limitations of Using Marine Plants for Synthesis of Nanoparticles	110
1.12	Research Gaps	111
1.13	Conclusion	111
References		112

Abstract Green synthesized nanoparticles from marine plants have gained significant demand as natural nanomedicines for drug delivery research, biosensing application, agro-industrial applications, biomedical industry, and textile industries. This book aimed to provide a timely comprehensive knowledge on different aspects of marine flora (seaweeds, seagrasses, mangroves, and other coastal plants)-derived nanoparticles (MFNPs). This comprehensive book has treatise more than twenty different applications of green synthesized metal (Ag, Au, Cd, Cu, Fe, Pd, Se) and metal oxide (CdO, CuO, Cu_2O, Cu/Cu_2O, Fe_3O_4, TiO_2, ZnO, Ag-ZnO, CdO-ZnO) nanoparticles derived from MFNPs. This book offers a detailed analysis of MFNPs, encompassing their synthesis, characterization, physiological properties, and regulatory factors. This book not only presents a comprehensive overview of MFNPs but also identifies key research gaps, outlining future directions for investigation. As such, it serves as a valuable resource for researchers and professionals in the field, offering actionable insights to propel advancements in MFNP research and applications. By providing a thorough understanding of MFNPs, this book aims to facilitate breakthroughs in the development and utilization of these innovative materials.

Keywords Nanoparticles · Phyto-nanomedicines · Marine flora · Nanodrug · Biocompatibility

1.1 Introduction

The increasing episodes of a wide variety of microbial infections and cancer types have raised a serious concern to clinicians and drug development researchers to search for novel molecules alternative to overused and indiscriminately misused drugs. Therefore, to treat and combat the rising incidents of new infections and diseases, and to bring nanotechnology based One Health approach, plant-based drugs, especially, plant derived nanoparticles have been identified as one of the alternative sources of novel nanodrugs in ethnobotanical studies. In ethnobiology, since ancient times, numerous plants were considered as most effective medicinal herbs which demonstrated to cure numerous diseases and illnesses. Thus, research focus has been shifted toward traditional medicinal plants to derive nanodrugs as potential bioactive drugs.

Multiple research studies have indicated the viability of various nanoparticles in the realm of nanomedicine, offering solutions for combating a plethora of microbial diseases and addressing the challenge of antimicrobial resistance. Although a variety of microbial (Bahrulolum et al. 2021; Hussain et al. 2016; Singh et al. 2018) and terrestrial plant species (Devatha and Thalla 2018; Jadoun et al. 2020), have widely been used to synthesize nanoparticles as effective nanodrugs, research on nanodrugs synthesis from marine plants is limited compared to terrestrial plants. However, marine plants originated nanoparticles have gained considerable interest due to promising bioactivity of nanodrugs synthesis through multiple mechanisms (Asmathunisha and Kathiresan 2013; Chaudhary et al. 2020; Fawcett et al. 2017)

1.1 Introduction

under various physicochemical factors (Jacob et al. 2021). Seaweeds are proven to have wide spectrum of biological activities due to numerous bioactive molecules (El-Beltagi et al. 2022; Safaat et al. 2021); thus, seaweeds (AlNadhari et al. 2021; El-Sheekh and El-Kassas 2016; Jeong et al. 2022; Puskulluoglu and Michalak 2022; Vidyasagar et al. 2023) and to some extent mangroves (Gouda et al. 2019), are used widely in green synthesis of nanoparticles to understand the enhanced bioactive nature of nanoparticles. Green synthesis of inorganic nanoparticles (e.g., copper, gold, iron, pallidium, platinum, silver, titanium, zinc, etc.) are gaining more interest due to their bioactive nature and their compatibility in drug delivery (Oscar et al. 2016). The green synthesized nanoparticles have a numerous beneficial application in fertilizers, cosmetics, textiles, antimicrobial agent in paint, bioremediation, electronics, wound healing, and drug delivery (Jacob et al. 2021; Roy 2019). This book provides a comprehensive literature on multiple applications of nanoparticles fabricated from marine plants and their usages and limitations.

1.1.1 Emphasis of Marine Plants in Ancient Literature

Since the ancient times, marine plants had been used in ancient medicinal history due to their potential bioactive molecules with different biological properties (Nag et al. 2022). Such important marine plants have been mentioned in the sacred scriptures such as the bible (Book of Jonah Chapter 2: Verse 5), Islamic literature (Khalilieh and Boulos 2006), and ancient and traditional literature (Pérez-Lloréns et al. 2023). Although ayurveda has been well recognized globally through India, the importance of marine plants in the ayurveda is not mentioned perhaps due to the main focus on terrestrial plants. China is one of the leading country in the world in exploring the applications of marine seaweeds as medicine since ages (Chengkui et al. 1984). Based on importance of medicinal properties of marine plants, researchers are exploring fabrication of nanoparticles to develop biocompatible and bioactive novel nanodrugs for the current age of drug resistance and new infections prevalence.

1.1.2 Benefits of Marine Flora-Derived Nanoparticles Over Terrestrial Plants

Unlike terrestrial plants, which obtain nutrients merely from soil, marine plants obtain a soup of nutrients and other chemical components regularly from seawater and possess potential phytochemicals like alkaloids, flavanoids, phenols, saponins, quinones, steroids, sterols, tannins, terpenoids, proteins, amino acids, and carbohydrates. Thus, search for novel drugs from marine plants is gaining more attention in the current research due to multiple biological properties of plant derived molecules.

Due to the medicinal properties of seawater chemistry, marine plants have been identified as potential sources for food and drug applications. Another advantage is that the derived nanoparticles are safe and display least cytotoxicity (Jacob et al. 2021). In addition, synthesizing nanoparticle in large quantities from fresh or dead macroalgal materials at extracellular level and its purification are quiet easier (Vijayaraghavan et al. 2011). Due to the innumerable applications demonstrated from various nanoparticles, particularly, silver nanoparticles, nanoparticles have gained more attention in electronics, textiles, medicine, and agriculture fields.

1.2 Literature Source and Search Strategy

This book was written based on published literature available from 2000 to 2023. Literature was accessed from online databases such as Google Scholar, PubMed, Scopus, Embase, ResearchGate, Web of Science, Science Direct, ProQuest, and Cochrane. Some inaccessible literature was obtained from the authors by personal requests. The primary search keywords used to obtain the literature data from the published articles of marine plants derived nanoparticles include "Nanoparticles", "Green synthesis", "biofabrication", "biosynthesis", "marine plants", "seaweed/macroalgae", "mangrove", "seagrass", and "nanostructure". In this book, nearly ~ 410 research articles reporting marine plants-based nanoparticles were covered to detail the various snippets of this book. All the graphical data reported in this book was derived from the primary tables presented within the text.

1.3 Advantages of Nanoparticles Derived from Marine Flora

Although various methods of synthesis of nanoparticles existed (Chugh et al. 2021; Salem and Fouda 2021; Dan Zhang et al. 2020a, b), nanoparticles originated from green synthetic pathways are gaining more interest in the recent decade due to the following reasons: (1) natural origin, (2) material (dead/fresh sample) availability, (3) availability of variety of reducing and capping agents, (4) eco-friendly, nontoxic, convenient, innocuous, and facile green synthesis, (5) one step process, (6) mostly extracellular production, (7) more stable, (8) large scale production, (9) cost-effective production by energy saving process, (10) effective drug properties, (11) biocompatibility (tested by hemolytic assay), (12) least agglomeration (aggregation) of NPs, (13) applications in a multiple way, (14) sustainable nanotechnology, and (15) controlled and targeted delivery to treat numerous diseases (Ahmad et al. 2019; Choudhary et al. 2022; Chugh et al. 2021; Salem and Fouda 2021; Shafey 2020; Vijayaraghavan et al. 2011). Thus, in recent years, green synthesized nanoparticles originated from plant sources are gaining more attention in the current research (Hussain et al. 2016; Jadoun et al. 2020; Shafey 2020).

1.4 Reducing and Capping Agents in Marine Flora

Marine plants possess potential phytochemicals that act as stabilizing, reducing, and capping agents, including alkaloids, flavanoids, polysaccharides, phenols, polyphenols, saponins, quinones, steroids, sterols, tannins, terpenoids, proteins, enzymes, fatty acids, amino acids, and carbohydrates (Fig. 1.1). The molecules present in the marine plants serve initially as reducing agents and then act as capping agents to form nanoparticles. Readers may refer to Sect. 1.5 (Table 1.1) for further information. The main functional groups in marine plant extracts such as CHO, COOH, OH, NH_2, SH, and other chemical components, act as reducing and stabilizing or capping agents (Choudhary et al. 2022; Chugh et al. 2021; Jacob et al. 2021). During the synthesis reaction, aqueous extracts of marine plants like seaweeds are reported to serve as potential reducing and capping agents (Algotiml et al. 2022b).

1.5 Green Synthesis of MFNPs

This section is specifically detailed for non-marine researchers to understand the sampling collection and nanoparticle synthesis (Figs. 1.2 and 1.3). The basic information that non-marine researchers must understand before going to the collection of marine plants are–(1) Low tide time: It is the best and most convenient time to collect exposed seaweeds and seagrasses easily by handpicking. In some places, seagrasses are found far away from shore; in this case, these plants need to be collected by snorkeling or SCUBA diving; (2) Crocodile free environment: Some mangrove regions in the world are inhabited by crocodiles. So, information on the distribution of such life-threatening organisms must be enquired and safety precautions must be taken before venturing into such study areas. Low tide time is always best and safer to collect mangrove samples. The tidal timetable can be checked at websites: www.tides4fishing.com, and www.tide-forecast.com; (3) Sample storage (most often marine plants are collected and placed in polythene zip lock bags), preservation (use of icebox is recommended to store the samples at lower temperatures to avoid plant decomposition and desiccation during a long transportation period or long period of sampling in the field. In case of well prepared and targeted sampling, plant materials are washed in the field itself, placed in cryocan containing liquid nitrogen, collected out and grounded in motor pestle to get powder form to make aqueous solution of marine plants) and transportation to the laboratory, and (4) Sample identification (it is imperative to identify the sample meticulously using standard taxonomic keys or molecular identification using *COI* or *ITS* gene barcodes).

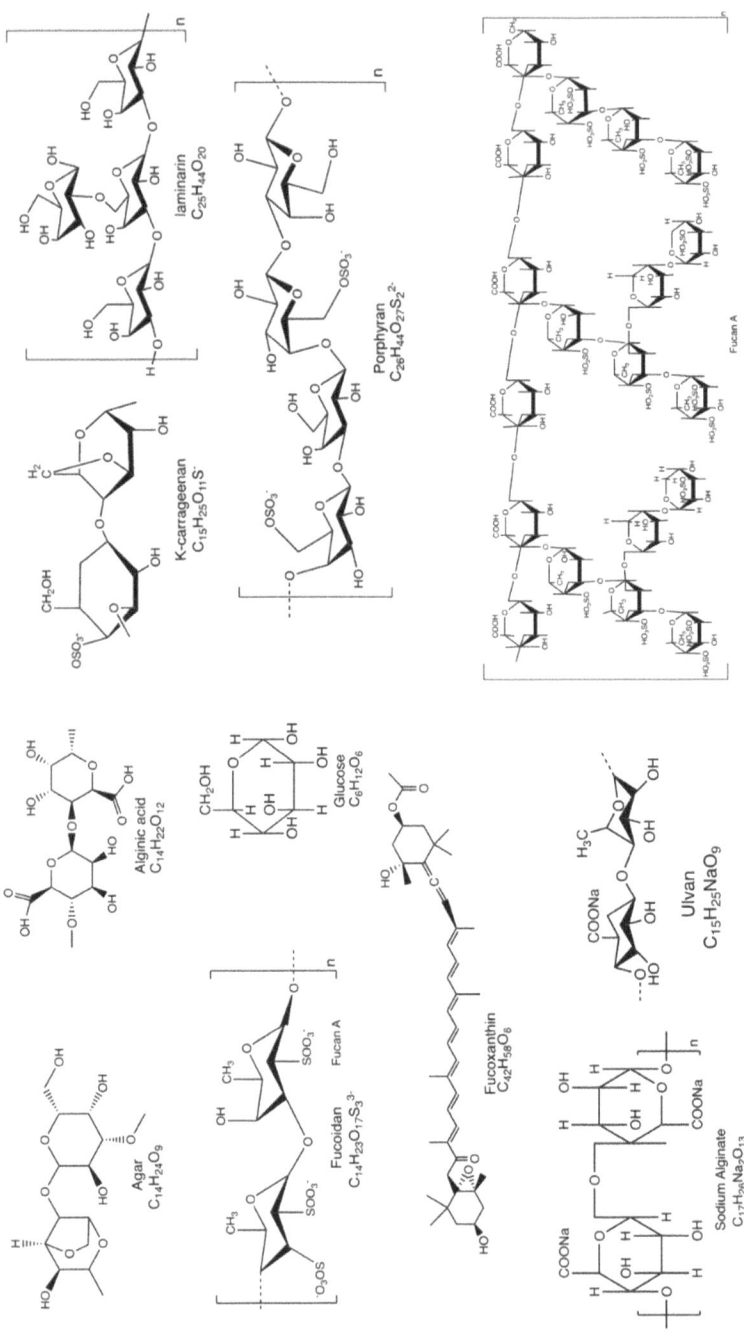

Fig. 1.1 Some reducing and capping agents originated from of marine plants

1.5 Green Synthesis of MFNPs

Table 1.1 Details of green synthesized nanoparticles from marine plants

Marine plant	*Plant part and concentration	Precursor concentration	Reaction time	Synthesis temperature	Reaction pH	Reaction mixture stirring	Synthesized nanoparticles	Shape	Size range	UV–VIS spectrum	Reference
Green seaweeds											
Caulerpa peltata	5 mL of DLE	95 mL of 1 mM zinc nitrate	1 h – 6 months	50–100 °C	5–10	Magnetic stirred	ZnONPs	Triangle, radial, hexagonal, rod, and rectangle	36–186 nm	372 nm	Nagarajan and Kuppusamy (2013)
Caulerpa prolifera	1 mL of DLE	99 mL of 10^{-3} M HAuCl$_4$	12 h	RT		120 rpm	AuNPs	Spherical	12.6–15.5 nm	555 nm	Kamal et al. (2022)
Caulerpa racemosa	10 of mL DLE	90 mL of 10^{-3} M AgNO$_3$	3 h	RT		–	AgNPs	Spherical with few triangular	5–25	413 nm	Kathiraven et al. (2015)
Caulerpa racemosa	5 mL of DLE	0.01 M AgNO$_3$	24 h	25 °C		–	AgNPs	Distorted spherical and few triangular	25 nm	441 nm	Edison et al. (2016)
Caulerpa scalpelliformis							AgNPs	Spherical	16.17–47.78 nm	415 nm	Manikandan et al. (2019)
Caulerpa scalpelliformis	FLE	1 mM AgNO$_3$		RT			AgNPs	Spherical and cubic	20–35 nm	350 nm	Murugan et al. (2015a, b)
Caulerpa serrulata	5–25 mL of DLE	10^{-3} M AgNO$_3$	1 h	95 °C	9.95		AgNPs	Spherical	10 ± 2 nm	412 nm	Aboelfetoh et al. (2017)
Caulerpa sertularioides							AgNPs	Spherical	24–57 nm	451 nm	Anjali et al. (2022)
Caulerpa taxifolia							SeNPs		28.6 nm		Men et al. (2009)
Caulerpa taxifolia	10 mL of purified DLE	100 mL of 4 mM AgNO$_3$	1 h	90 °C	–	Stirred	AgNPs	Spherical	10–100 nm	420 nm	Danjie Zhang et al. (2020a, b)
Chaetomorpha antennina	50 mL of DLE	50 mL of 1 mM AgNO$_3$	1 h	60 °C			AgNPs	Spherical	69.99–99.19 nm	452 nm	Kingslin et al. (2022a, b)

(continued)

Table 1.1 (continued)

Marine plant	*Plant part and concentration	Precursor concentration	Reaction time	Synthesis temperature	Reaction pH	Reaction mixture stirring	Synthesized nanoparticles	Shape	Size range	UV–VIS spectrum	Reference
Chaetomorpha antennina							AgNPs	Spherical	24 ± 2.4 nm	420 nm	Thanigaivel et al. (2021)
Chaetomorpha antennina	10 mL of WE	90 mL of 1 mM AgNO$_3$	72 h	RT	–	120 rpm	AgNPs	Hexagonal and cubical	256.2 nm	435.5 nm	Roy and Anantharaman (2017)
Chaetomorpha linum	20 mL of DLE	100 mL of 1 mM AgNO$_3$	30 min	37 °C	–	Static	AgNPs	Nano-clusters coalescence	3–44 nm	422 nm	Ragupathi Raja Kannan et al. (2013a, b)
Chlorodesmis hildebrandtii					–		AgNPs	Spherical	10–100 nm	448 nm	Roy and Anantharaman (2018a)
Cladophora glomerata	10 mL of DLE	90 mL of 1 mM AgNO$_3$	24 h	–	–	–	AgNPs	Irregular and sponge-like	200 nm	443 nm	Minhas et al. (2018)
Codium capitatum	12 mL of DLF and FLE	88 mL of 1 mM AgNO$_3$		RT		Static	AgNPs	Nano-clusters	3–44 nm	422 and 425 nm	Kannan et al. (2013a, b)
Codium tomentosum	FLE	1 mM AgNO$_3$	120 min	RT	–	–	AgNPs	Irregular	20–40 nm	420 nm	Murugan et al. (2016b)
Codium tomentosum	1 gm/mL FLE	0.4 mM HAuCl$_4$·3 H$_2$O	24 h	RT	–	Stirring	AuNPs	Spherical	34.5 ± 5.6 nm	539 nm	González-Ballesteros et al. (2023)
Enteromorpha compressa	DLE	90 mL of 1 mM AgNO$_3$	1 h	RT			AgNPs	Spherical	4–24 nm	421 nm	Vijayan Sri Ramkumar et al. (2017a, b)
Enteromorpha compressa	1 mL of DLE	99 mL of 1 mM AgNO$_3$	12 h	RT			AgNPs	Spherical	40–50 nm	420 nm	Dhanalakshmi et al. (2012)
Enteromorpha flexuosa	10 mL of DLE	90 mL of 1 mM AgNO$_3$	60 min	–	–	Dark room	AgNPs	Spherical	2–32 nm	430 nm	Yousefzadi et al. (2014)
Enteromorpha flexuosa	50 mL of DLE	450 mL of 10^{-3} M AgNO$_3$	60 min	RT	–	Dark room	AgNPs	Spherical	30–90 nm	434 nm	Azeem et al. (2022)
Enteromorpha intestinalis	20 mL of DLE	80 mL of 1 mM AgNO$_3$	24 h	RT	–	Stirring	AgNPs	Spherical and cubic	10–20 nm	436 nm	Raju et al. (2017)

(continued)

1.5 Green Synthesis of MFNPs

Table 1.1 (continued)

Marine plant	*Plant part and concentration	Precursor concentration	Reaction time	Synthesis temperature	Reaction pH	Reaction mixture stirring	Synthesized nanoparticles	Shape	Size range	UV–VIS spectrum	Reference
Enteromorpha linza	5 mL of DLE	25 mL of 3 mM AgNO$_3$	24 h	RT			AgNPs	Spherical		426 nm	Muthu et al. (2014)
Enteromorpha prolifera							AgNPs	Polydispersed and spherical	17.8 nm	448 nm	Kingslin et al. (2022a)
Halimeda gracilis							AgNPs	Cubical, hexagonal and irregular shaped	< 100 nm	430 nm	Roy and Anantharaman (2018b)
Halimeda opuntia	10 mL of DLE	90 mL of 30 mM selenious acid and 1.8 mL of 40 mM ascorbic acid	2 h	RT	–	160–170 rpm	SeNPs	Spherical	30–80 nm	294 nm	Radhika et al. (2022)
Rhizoclonium fontinale	DLE	HAuCl$_4$	72 h	–	9	–	AuNPs	Spherical	~16 nm	530 nm	Parial et al. (2012)
Rhizoclonium hieroglyphicum	WE	198 Au radiotracer solution	72 h	–	8	–	AuNPs	Spherical	< 20 nm	–	Chakraborty et al. (2009)
Ulva armoricana	10 gm of FLE	100 mL of HAuCl$_4$·3 H$_2$O	7 days	Sunlight			AuNPs	Spherical, triangular and rod	200 nm	550 nm	Mukhoro et al. (2018)
Ulva armoricana	Ulvan	AgNO$_3$		RT		Stirring	AgNPs	Spherical	12.5 nm	416 nm	Massironi et al. (2019)
Ulva compressa	10 mL of DLE	90 mL of 1 mM AgNO$_3$	24 h	–	–	–	AgNPs	Irregular and sponge-like	200 nm	443 nm	Minhas et al. (2018)
Ulva fasciata							AgNPs	Spherical	4–80 nm	430 nm	Negm et al. (2018)
Ulva fasciata	10 mL of DLE	90 mL of 0.1 mM AgNO$_3$	–	40 °C	–	Stirring	AgNPs	Spherical	9–37 nm	430 nm	Alshubaily et al. (2020)
Ulva fasciata	3 mL of 0.1% *U. fasciata* ethyl acetate extract	100 mL of 10^{-3} M AgNO$_3$		RT			AgNPs	Spherical and poly-dispersed	28–41 nm	440 nm	Rajesh et al. (2012)

(continued)

Table 1.1 (continued)

Marine plant	*Plant part and concentration	Precursor concentration	Reaction time	Synthesis temperature	Reaction pH	Reaction mixture stirring	Synthesized nanoparticles	Shape	Size range	UV–VIS spectrum	Reference
Ulva faciata	90 mL of crude hot water soluble polysaccharide solution	1 mL of 0.1 mM $AgNO_3$	20 min	70 °C	10	Stirring	AgNPs	Spherical	7 nm	407–424 nm	El-Rafie et al. (2013)
Ulva fasciata	10 mL of DLE	190 mL of 1 mM $HAuCl_4$	20 min	RT	–	–	AuNPs	Spherical	10 ± 3 nm	541 nm	Kumari et al. (2014)
Ulva fasciata	50 mL of DLE	50 mL of 1 mM $AgNO_3$	12 h	RT		Stirring	Ag/AgCl NPs	Spherical and few polygonal	38.44–58.17 nm	413.5 nm	Lashgarian et al. (2021)
Ulva fasciata	2 mL of DLE	10 mL of 10 mM $Zn(CH_3CO_2)_2 \cdot 2H_2O$	4 h	72 °C		Stirring	ZnONPs	Flower and sphere	77.81 nm		Alsagaf et al. (2021)
Ulva fasciata	10 mL of DLE	90 mL of 0.1 mM $AgNO_3$	–	40 °C	–	Stirring	AgNPs	Spherical	30–500 nm	426–446 nm	Hamouda et al. (2018)
Ulva fasciata							SeNPs		20–50 nm		Shahzamani et al. (2022)
Ulva fasciata	10 mL of DLE	90 mL of 1 mM $AgNO_3$	40 min	30 °C	–	Magnetic stirring	AgNPs	–	–	428 nm	Khalifa et al. (2016)
Ulva fasciata	DLE	0.1 M $FeCl_3$	1 h 30 min	RT	–	800 rpm	Fe_3O_4 NPs	Nanospheres	22.73–39.77 nm	400 nm	Salem et al. (2020)
Ulva fasciata	10 mL of DLE	90 mL of 2 mM zinc acetate	60 min	50 °C	–	Stirring	ZnONPs	Spherical	3–33 nm	330 nm	Fouda et al. (2022b)
Ulva flexuosa	10 mL of DLE	90 mL of 1 mM $AgNO_3$	60 min	RT	–	–	AgNPs	Spherical	2–32 nm	430 nm	Rahimi et al. (2014)
Ulva flexuosa	DLE	0.10 M Fe^{3+}/ Fe^{2+} chloride solution	10 min	88 °C	8–11	Magnetic stirring	FeNPs	Cubo-spherical crystalline	10–15 nm	–	Mashjoor et al. (2018)
Ulva flexuosa	1% DLE	Copolymer beads were equilibrated with 0.01 M $AgNO_3$	24 h	25 °C	–	Ultrasonic irradiation & 120 rpm	AgNPs	Cubic	4.93–6.70 nm	480 nm	Dixit et al. (2018)
Ulva intestinalis	WE	15 ppm $HAuCl_4$	72 h	20 °C			AuNPs	Spherical to irregular	~42.39 nm	542 nm	Parial et al. (2012)

(continued)

1.5 Green Synthesis of MFNPs

Table 1.1 (continued)

Marine plant	*Plant part and concentration	Precursor concentration	Reaction time	Synthesis temperature	Reaction pH	Reaction mixture stirring	Synthesized nanoparticles	Shape	Size range	UV–VIS spectrum	Reference
Ulva intestinalis	1 g/mL of FLE	0.16 mM of AgNO$_3$	1 h	100 °C	6.40	Stirring	AgNPs	Spherical	14.2 ± 2.0 nm	409 nm	González-Ballesteros et al. (2019a, b)
Ulva intestinalis	1 g/mL of FLE	0.5 mM HAuCl$_4$	24 h	RT	4.73	Stirring	AuNPs	Spherical	17.8 ± 2.7 nm	531 nm	González-Ballesteros et al. (2019a, b)
Ulva lactuca	0.5 g/mL FLE	0.17 mM AgNO$_3$	8 h	RT	6.12	–	AgNPs	Spherical	31 ± 8 nm	423 nm	González-Ballesteros et al. (2019a, b)
Ulva lactuca	1 g/mL FLE	0.4 mM HAuCl$_4$	24 h	RT	4.28	–	AuNPs	Spherical	7.9 ± 1.7 nm	529 nm	González-Ballesteros et al. (2019a, b)
Ulva lactuca	10 mL of DLE	1 mM AgNO$_3$	24 h	100° C	–	Stirring	AgNPs	Spherical	9–30 nm	425 nm	Amin (2020)
Ulva lactuca	100 mL of DLE	1 mM AgNO$_3$					AgNPs	Spherical	20–50 nm	453 nm	Aziz (2022)
Ulva lactuca	50 mL of FLE	50 mL of 1 mM AgNO$_3$		RT			AgNPs	Spherical	48.59 nm	430 nm	Kumar et al. (2013a)
Ulva lactuca	1 mL of Ulvan filtrate	10 mL of 30 mM of selenous acid solution and 200 mL of 40 mM ascorbic acid				250 rpm	SeNPs	Spherical	85 nm	270 nm	Vikneshan et al. (2020)
Ulva lactuca	1 gm of DLE	1 mM AgNO$_3$		55 °C	11		AgNPs	Spherical aggregated	40.19 ± 5.93	420–450 nm	Koçer and Özçimen (2022)
Ulva lactuca	DLE	1 mM AgNO$_3$	10 min	121 °C		Magnetic stirring	AgNPs	Spherical	20–56 nm	434 nm	Devi and Bhimba (2012)
Ulva lactuca	DLE						AgNPs				Bhimba and Kumari (2014)
Ulva lactuca	DLE	1 mM AgNO$_3$	–	RT	–	–	AgNPs	Cubical	20–35 nm	400 nm	Murugan et al. (2015b)

(continued)

Table 1.1 (continued)

Marine plant	*Plant part and concentration	Precursor concentration	Reaction time	Synthesis temperature	Reaction pH	Reaction mixture stirring	Synthesized nanoparticles	Shape	Size range	UV-VIS spectrum	Reference
Ulva lactuca	5 mL of DLE	95 mL of 1 mM zinc acetate	4 h	70 °C	–	Magnetic stirring	ZnONPs	Asymmetrical	10–50 nm	325 nm	Ishwarya et al. (2018a)
Ulva lactuca	5 mL of DLE	95 mL of 10^{-3} M AgNO$_3$	–	–	8.3	–	AgNPs	Spherical, hexagonal, oval and triangular	22.65 ± 3.44 nm	411 nm	Sahayaraj et al. (2018)
Ulva lactuca	DLE	1 mM AgNO$_3$	2 h	RT	–	Stirring	AgNPs	Spherical	~ 24 nm	440 nm	Gurusamy et al. (2019)
Ulva lactuca	10 mg of DLE	4 mL of 10 mM NaCl + methanol + 100 mL of glutaraldehyde (10%)	24 h	4 °C	7.5	Stirring	Ulva lactuca loaded albumin nanoparticles (ULANP)	–	–	–	Al-Malki (2020)
Ulva lactuca	1 mg/mL FLE	100 mL of 5 mM AgNO$_3$	–	25 °C	–	–	AgNPs	Spherical	8–14 nm	452 nm	Acharya et al. (2022)
Ulva lactuca	25 mL of DLE	25 mL of 0.1 M FeCl$_3$	2 h	RT	–	–	FeNPs	Spherical	30–40 nm	275–325 nm	Bensy et al. (2022)
Ulva reticulata	1 mL of DLE	99 mL of 1 mM AgNO$_3$	12 h	RT			AgNPs	Spherical	40–50 nm	420 nm	Dhanalakshmi et al. (2012)
Ulva reticulata	5 gm of DLE	1 mM AgNO$_3$	10 min	121 °C			AgNPs	Spherical	20–50 nm	410–440 nm	Bhimba and Devi (2014)
Ulva rigida	20 mL of DLE	80 mL of 10^{-3} M AgNO$_3$	15 min	70 °C	–	Stirred on a hot plate	AgNPs	Spherical	12 nm	424 nm	Algotiml et al. (2022b)
Ulva rigida	WE	10^{-3} M	–	RT	–	HM	AuNPs	Spherical	9 nm	528 nm	Algotiml et al. (2022a)
Urospora sp.	DLE	1 mM AgNO$_3$	30 min	70 °C		Magnetic stirring	AgNPs	Spherical	20–30 nm	430 nm	Suriya et al. (2012)
Valoniopsis pachynema	5 mL of FLE	50 mL of 2 mM cadmium chloride	12 h	30 °C		200 rpm	CdSNPs	Spherical	< 100 nm	295 nm	Sujitha et al. (2017)
Valoniopsis pachynema	DLE	10^3 M AgNO$_3$					AgNPs		30–40 nm	408 nm	Selvaraj et al. (2020)

(continued)

1.5 Green Synthesis of MFNPs

Table 1.1 (continued)

Marine plant	*Plant part and concentration	Precursor concentration	Reaction time	Synthesis temperature	Reaction pH	Reaction mixture stirring	Synthesized nanoparticles	Shape	Size range	UV-VIS spectrum	Reference
Brown Seaweeds											
Brown alga	Alginic acid	Pd					PdNPs		2–20 nm		Parker et al. (2015)
Bifurcaria bifurcata	2 mL of DLE	20 mL of 1 mM copper(II) sulfate	24 h	100–120 °C		Stirring	CuONPs	Spherical	5–45 nm	260 nm	Abboud et al. (2014)
Cladosiphon okamuranus	Fucoidan	3.45 μL of $HAuCl_4 \cdot 3H_2O$	45 min	85 °C	–	Stirring	AuNPs	Spherical	8–10 nm	527–530 nm	Soisuwan et al. (2010)
Colpomenia sinuosa	90 mL of crude hot water soluble polysaccharide solution	1 mL of 0.1 mM $AgNO_3$	20 min	70 °C	10	Stirring	AgNPs	Spherical	20 nm	407–424 nm	El-Rafie et al. (2013)
Colpomenia sinuosa	DLE	0.1 M $FeCl_3$	1 h	RT		800 rpm	Fe_3O_4 NPs	Nanospheres	11.24–33.71 nm	402 nm	Salem et al. (2019)
Colpomenia sinuosa	500 mg of DLE	10^{-3} M $AgNO_3$	24 h	RT	5.09	Stirring	AgNPs	Cubical	54–65 nm	420 nm	Kiran and Murugesan (2014a)
Colpomenia sinuosa	–	–	–	–	–	–	AgNPs	–	–	–	Vishnu and Murugesan (2014)
Cystophora moniliformis	10 mL of DLE	90 mL of 1 mM $AgNO_3$	30 min				AgNPs	Spherical	50–100 nm	Temperature dependent	Prasad et al. (2013)
Cystoseira baccata	1 mL of FLE	75 μL of 0.01 M $HAuCl_4$	24 h	RT		Stirring	AuNPs	Spherical and polycrystalline	8.4 ± 2.2 nm	532 nm	González-Ballesteros et al. (2017)
Cystoseira crinita	5 mL of DLE	15 mL of 0.05 M $ZnSO_4$	30 min	45 °C	6.5	Stirring	ZnONPs	Multilayered rectangular	23–200 nm	268 nm and 360	Elrefaey et al. (2022)
Cystoseira crinita	90 mL of DLE	10 mL of 2 mM $Mg(NO_3)_2 \cdot 6H_2O$	24 h	–	–	–	MgONPs	Spherical	3–18 nm	320 nm	Fouda et al. (2022b)
Cystoseira myrica	WE	10^{-3} M	–	RT	–	HM	AuNPs	Spherical	11 nm	540 nm	Algotiml et al. (2022a)
Cystoseira myrica	20 mL of DLE	80 mL of 10^{-3} M $AgNO_3$	15 min	70 °C	–	Stirred on a hot plate	AgNPs	Spherical	17 nm	409 nm	Algotiml et al. (2022b)
Cystoseira myrica							AgNPs	Polydisperse and spherical	8–15 nm	434 nm	Mohamed et al. (2022)

(continued)

Table 1.1 (continued)

Marine plant	*Plant part and concentration	Precursor concentration	Reaction time	Synthesis temperature	Reaction pH	Reaction mixture stirring	Synthesized nanoparticles	Shape	Size range	UV–VIS spectrum	Reference
Cystoseira myrica	1 mL of DLE	99 mL of 10^{-3} M HAuCl$_4$	12 h	RT		120 rpm	AuNPs	Spherical	12.6–15.5 nm	545 nm	Kamal et al. (2022)
Cystoseira myrica	10 mL of DLE	90 mL of 1 mM AgNO$_3$	40 min	30 °C	–	Magnetic stirring	AgNPs	–	–	428 nm	Khalifa et al. (2016)
Cystoseira trinodis	1 mL of DLE	99 mL of 10^{-3} M HAuCl$_4$	12 h	RT		120 rpm	AuNPs	Spherical	12.6–15.5 nm	540 nm	Kamal et al. (2022)
Cystoseira trinodis	2 mL of DLE	20 mL of 1 mM copper(II) sulfate	90 min			Mixed under ultrasonic condition (400w, 20kH)	CuO NPs	Spherical	6–7.8 nm	330 nm	Gu et al. (2018)
Desmarestia antarctica	1 gm/mL of FLE	0.4 mM HAuCl$_4$	15 h	RT	5.05	Stirring	AuNPs	Spherical	12.6 ± 1.9 nm	532 nm	González-Ballesteros et al. (2021)
Desmarestia antarctica	0.25 gm/mL of FLE	0.17 mM AgNO$_3$	1 h	100 °C	5.96	Stirring	AgNPs	Spherical	13.7 ± 3.1 nm	417 nm	González-Ballesteros et al. (2021)
Desmarestia menziesii	FLE	50 and 150 μL of 0.01 M HAuCl$_4$ and 100–500 μL of 0.005 M AgNO$_3$		RT and varied temperatures		Stirring	AuNPs and AgNPs	Spherical	11.5 ± 3.3 nm and 17.8 ± 2.6 nm	527 nm and 405 nm	González-Ballesteros et al. (2018)
Dictyopteris divaricata	2 mg/mL of DLE	1 mM HAuCl$_4$	10 min	22–25 °C	–	Stirring	AuNPs	Spherical and hexagonal	28.01 ± 2.03 nm	525 nm	Park et al. (2019)
Dictyota bartayresiana	10 mL of DLE	90 mL of 1 mM HAuCl$_4$	45 min	RT	–	–	AuNPs	Spherical	–	548 to 564	Varun et al. (2014)
Dictyota bartayresiana	10 gm of FLE	1 mM AgNO$_3$					AgNPs			410 nm	Antonysamy et al. (2015)
Ecklonia cava	1 mL of DLE	10 mL of 1 mM HAuCl$_4$	10 min	80 °C	–	–	AuNPs	Spherical and triangular	30 ± 0.25 nm	532 nm	Venkatesan et al. (2014)
Ecklonia cava	10 mL of DLE	90 mL of 1 mM AgNO$_3$	72 h	RT	–	Stirring	AgNPs	Spherical	43 nm	418 nm	Venkatesan et al. (2016)
Ecklonia stolonifera	2 mg/mL of DLE	1 mM HAuCl$_4$·3H$_2$O	15 min	25 °C	–	Stirring	AuNPs	Spherical and hexagonal	27.9 ± 4.3 nm	543 nm	Jun et al. (2020)

(continued)

1.5 Green Synthesis of MFNPs

Table 1.1 (continued)

Marine plant	*Plant part and concentration	Precursor concentration	Reaction time	Synthesis temperature	Reaction pH	Reaction mixture stirring	Synthesized nanoparticles	Shape	Size range	UV–VIS spectrum	Reference
Egregia sp.	2.5 mg/ml of DLE	0.1 mM HAuCl$_4$	10 min	100 °C	–	–	AuNPs	Spherical	8 nm	530 nm	Colin et al. (2018)
Fucus evanescens	10 mg of alginate and fucoidan	100 mL of 1 mM AgNO$_3$	30 min	RT	10–11	Stirring	AgNPs	Spherical	57 ± 9 – 64 ± 6 nm	402–408 nm	Yugay et al. (2020)
Fucus vesiculosus	1 gm/L of DLE	75 mL of 100 mg/L HAuCl$_4$	8 h	RT	7	Stirring	AuNPs	Spherical and variety of shapes	5–50 nm	–	Mata et al. (2009)
Fucus vesiculosus	100 mg of acetylated fucoidan	DOX solution	Overnight	–	–	Stirring	DOX-loaded acetylated fucoidan nanoparticles (DOX-AcFuNPs)	Spherical	140 nm	–	Lee et al. (2013)
Fucus vesiculosus	3 mL of 0.5 mg/ml of fucoidan solution	3 mL of 0.5 mg/ml of chitosan solution	–	–	9	–	Fu/CHNPs	Round	115–140 nm	–	Oliveira et al. (2018)
Fucus vesiculosus	Fucoidan solution	DOX + PEI solution	–	–	–	–	PEI-FCD-DOX NPs	Spherical	41.1 ± 17.39 –159.2 ± 18.34 nm	–	Pawar et al. (2019)
Fucus vesiculosus	5 mg of Fucoidan	1 × 10^{-4} M HAuCl$_4$.3H$_2$O	30 min	80 °C	7.2	Stirring	AuNPs	Spherical	15 –119 nm	570 nm	Khan et al. (2019)
Halopteris scoparia	1 gm of DLE	1 mM AgNO$_3$		55 °C	11		AgNPs	Spherical aggregated	34.88 ± 4.38	420–450 nm	Koçer and Özçimen (2022)
Kjellmaniella crassifolia	Fucoidan	3.45 μL of HAuCl$_4$.3H$_2$O	45 min	85 °C	–	Stirring	AuNPs	Spherical	8–10 nm	527–530 nm	Soisuwan et al. (2010)
Laminaria digitata	20 mL of 10 g/L laminarin	100 mL of 2 M zinc acetate dehydrate	6 h	–	–	Stirring	ZnONPs	Spindle	100–350 nm	382.25 nm	Vijayakumar et al. (2022)
Laminaria japonica	Fucoidan solution (1.0–5.0 mg/mL, 1 mL)	Protamine solution (1.0–5.0 mg/mL, 1 mL)	2 min			Stirring	Protamine/fucoidan NPs	Spherical	180 nm	198 nm and 217 nm	Lu et al. (2017)
Laminaria japonica	2 mL of DLE	20 mL of 2 mM AgNO$_3$					AgNPs	Spherical- to oval	20–31 nm	405 nm	Kim et al. (2018)
Laminaria japonica							AuNPs	Spherical	15–20 nm		Ghodake and Lee (2011)

(continued)

Table 1.1 (continued)

Marine plant	*Plant part and concentration	Precursor concentration	Reaction time	Synthesis temperature	Reaction pH	Reaction mixture stirring	Synthesized nanoparticles	Shape	Size range	UV-VIS spectrum	Reference
Laminaria ochroleuca	1.5% sodium alginate solution	0.05 M AgNO$_3$	10 min	60 °C	–	Stirring	AgNPs	Spherical	10–20 nm	400 nm	Kaidi et al. (2022)
Lobophora variegata	10 mL of FLE	90 mL of 1 mM AgNO$_3$		RT		Mechanical stirring	AgNPs	Spherical	20–50 nm	420 nm	Sathyaseelan et al. (2015)
Padina sp.	20 mL of DLE	180 mL of 0.01 M AgNO$_3$	48 h	60 °C	–	Stirring	AgNPs	Spherical, oval-shaped, irregular and polydis- persed	25–60 nm	420 nm–445 nm	Bhuyar et al. (2020)
Padina australis	5 mL of DLE	95 mL of 1 mM AgNO$_3$	1 h	60 °C	–	–	AgNPs	Spherical	2.5–23 nm	421 nm	Kailasam et al. (2023)
Padina boeregeseni	20 mL of DLE	100 mL of 1 mM AgNO$_3$	24 h	RT	8.5	–	AgNPs	Spherical	34.62–54.33 nm	418 nm	Hashemi et al. (2015)
Padina boryana	5 mL of DLE	100 mL of Na$_2$PdCl$_4$	2 h	60 °C	–	200 rpm	PdNPs	Spherical	5–20 nm	293 nm	Sonbol et al. (2021)
Padina gymnospora	10 mL of DLE	190 mL of 1 mM HAuCl$_4$	12 h	75 °C	10	–	AuNPs	Spherical	53–67 nm	527 nm	Singh et al. (2013a, b)
Padina gymnospora	10 mL of DLE	40 mL of 0.004 M of AgNO$_3$	–	30 °C	–	120 rpm	AgNPs	Spherical	25–40 nm	410 nm	Shiny et al. (2013a, b)
Padina gymnospora	10 mL of DLE	20 mL of 0.001 M H$_2$PtCl$_6$		50 °C		Intermittent shaking	PtNPs	Spherical	5–20 nm	260–280 nm	Shiny et al. (2014)
Padina gymnospora	FLE	1 mM HAuCl$_4$		75 °C	9.5		AuNPs	Spherical	8–21 nm	544 nm	Singh et al. (2014)
Padina gymnospora	FLE	10^{-3} M HAuCl$_4$	10 min	30–95° C	3.5 to 11.5	Water bath shaker	AuNPs	Spherical, triangles, pentagons and hexagons	14 nm	520 nm	Singh et al. (2015)
Padina gymnospora	10 l of DLE	90 mL of 0.001 M platinum chloride (H$_2$PtCl$_6$)	10–15 min	–	–	Stirring	PtNPs	Truncated octahedral	5–50 nm	370 nm	Sri Ramkumar et al. (2017a, b)

(continued)

1.5 Green Synthesis of MFNPs

Table 1.1 (continued)

Marine plant	*Plant part and concentration	Precursor concentration	Reaction time	Synthesis temperature	Reaction pH	Reaction mixture stirring	Synthesized nanoparticles	Shape	Size range	UV–VIS spectrum	Reference
Padina gymnospora	0.5 gm of DLE	0.25 M ZnCl (25 ml) solution and 0.1 M CdCl (25 ml)	4–24 h	RT	–	Stirring	CdO-ZnONPs	Small fibrillar pattern	20–50 nm	237 nm and 370 nm	Rajaboopathi and Thambidurai (2017)
Padina pavonica	1 gm of DLE	100 mL of 10^{-3} M AgNO$_3$	3 h	RT	–	Stirring	AgNPs	Spherical and polygonal	49.58–86.37 nm	401 nm	Abdel-Raouf et al. (2019)
Padina pavonica	75 mL of DLE	0.19 gm of 10^{-3} M HAuCl$_4$ in 450 ml distilled water	24 h	RT	–	–	AuNPs	Spherical	30–100 nm	545.5 nm	Isaac and Renitta (2015)
Padina pavonica	10 mL of DLE	90 mL of 10^{-3} mM AgNO$_3$	24 h				AgNPs	Spherical polydispersed	45–54 nm	422 nm	Sahayaraj et al. (2012)
Padina pavonica	DLE	0.1 mol/L of FeCl$_3$	60 min	RT	–	Stirring	Fe$_3$O$_4$NPs	Spherical	10–19.5 nm	402 nm	El-Kassas et al. (2016)
Padina pavonica	10 mL of DLE	90 mL of 10^{-3} mM AgNO$_3$	Few hours	RT	–	–	AgNPs	Spherical	20–70 nm	422 nm	Sudha and Balasundaram (2018)
Padina tetrastromatica	0.5 mg of sodium alginate	100 mL of 1 mM AgNO$_3$	60 min	80 °C	8–9	–	AgNPs	Spherical and small aggregates	54–78 nm	417 nm	Sangeetha et al. (2012)
Padina tetrastromatica	FLE	0.1, 0.25, 0.5, 0.75, 1 mM of AgNO$_3$	48 h	37 °C	–	–	AgNPs	Cubical	18–20 nm	450 nm	Kayalvizhi et al. (2014)
Padina tetrastromatica	FLE	0.1, 0.25, 0.5, 0.75, 1 mM of HAuCl$_4$	48 h	37 °C	–	–	AuNPs	Cubical	20–90 nm	550 nm	Kayalvizhi et al. (2014)
Padina tetrastromatica				60 °C & RT			AgNPs	Spherical	4–35 nm	420 nm	Shiny et al. (2013a, b)
Padina tetrastromatica	50 mL of DLE	50 mL of 1 mM AgNO$_3$	1 h		–	Static	AgNPs	Spherical	85.85–89.75 nm	422 nm	Kingslin and Ravikumar (2016)
Padina tetrastromatica	10 mL of fucoidan	90 mL of 1 mM AgNO$_3$	24 h	–	–	Stirring	AgNPs	Spherical	4–25 nm	440 nm	Rajeshkumar (2017)

(continued)

Table 1.1 (continued)

Marine plant	*Plant part and concentration	Precursor concentration	Reaction time	Synthesis temperature	Reaction pH	Reaction mixture stirring	Synthesized nanoparticles	Shape	Size range	UV–VIS spectrum	Reference
Padina tetrastromatica	DLE	1 mM HAuCl$_4$	48 h	20–80 °C	6–9		AuNPs	Spherical	8–10 nm	527 nm	Rajeshkumar et al. (2017a)
Padina tetrastromatica	10 mL of FLE	90 mL of 1 mM AgNO$_3$	72 h	300 °C		120 rpm	AgNPs	Spherical	10–100 nm	426 nm	Jegadeeswaran et al. (2012)
Padina tetrastromatica							AgNPs	Spherical	14 nm	460 nm	Rajeshkumar et al. (2012a, b)
Padina tetrastromatica	50 mL of FLE	50 mL of 1 mM AgNO$_3$	1 h	60 °C	–	Stirring	AgNPs	Spherical	5–35 nm	424 nm	Princy and Gopinath (2015)
Padina tetrastromatica	10 mL of DLE	1 mM AgNO$_3$	72 h	37 °C	–	Stirring	AgNPs	Round	40–50 nm	430 nm	Selvi et al. (2016)
Padina tetrastromatica	10 mL of DLE	90 mL of 1 mM AgNO$_3$	24 h	RT	–	Stirring	AgNPs	Spherical, cubical, truncated, and ellipsoidal	–	440 nm	Rajeshkumar et al. (2017b)
Padina tetrastromatica							ZnONPs				Rajeshkumar (2018)
Polycladia myrica	DLE	1 mM Na$_2$SeO$_4$	72 h	25.2 °C	–	Stirring	SeNPs	Oval, spherical, and semispherical	9.31–68.65 nm	350 nm	Touliabah et al. (2022)
Polycladia myrica	DLE	1 mM Na$_2$SeO$_4$	72 h	25.2 °C	–	Stirring	SeNPs	Spherical	17.48–23.01 nm	350 nm	Abo-Neima et al. (2023)
Rosenvingea intricata	50 mL of FLE	0.6 M FeCl$_3$.6H$_2$O	1 h	80 °C	11	Stirring	Fe$_3$O$_4$NPs	Irregular, spherical	14.36–125 nm	400–550 nm	Sri et al. (2024)
Saccharina cichorioides	10 mg of fucoidan and laminaran	100 mL of 1 mM AgNO$_3$	30 min	RT	10–11	Stirring	AgNPs	Spherical	45 ± 2 – 53 ± 4 nm	412–414 nm	Yugay et al. (2020)
Saccharina japonica	50 mL of DLE	50 mL of 1 mM AgNO$_3$	45 min	40 °C	–	200 rpm	AgNPs	Spherical	14.77 nm	420 nm	Sivagnanam et al. (2017)
Sargassum sp.	5–10 mL of DLE	1–2 mL of 0.01 M HAuCl$_4$	48 h	RT		Stirring	AuNPs	Triangular, truncated triangular, and hexagonal	5–400 nm	525 nm	Liu et al. (2005)

(continued)

1.5 Green Synthesis of MFNPs

Table 1.1 (continued)

Marine plant	*Plant part and concentration	Precursor concentration	Reaction time	Synthesis temperature	Reaction pH	Reaction mixture stirring	Synthesized nanoparticles	Shape	Size range	UV–VIS spectrum	Reference
Sargassum sp.	DLE	95 mL of 0.1 M zinc nitrate	3 h	–	–	Stirring	ZnO NPs	Spherical	53.75–734.65 nm	–	Sari et al. (2017)
Sargassum sp.	20 gm of DLE	200 mL of 0.1 mol nickel chloride and 0.1 mol ferric chloride solution	8 h	–	–	Stirring	Ni-Fe NPs	Spherical	8.1–10.2 nm	–	Liang et al. (2020)
Sargassum acinarium	DLE	0.1 mol/L of FeCl$_3$	60 min	RT	–	Stirring	Fe$_3$O$_4$ NPs	Spherical	21.6–27.4 nm	415 nm	El-Kassas et al. (2016)
Sargassum angustifolium	DLE	1 mM AgNO$_3$	90 min	70 °C	–	Stirring	AgNPs	Spherical	32 ± 10 nm	428 nm	Ghaemi and Gholamipour (2017)
Sargassum angustifolium	5 mg of DLE	Selenious acid (1, 5, and 10 mM) and a stock solution of ascorbic acid (40 mM)	–	40 °C	–	Stirring	Algae coated SeNPs	Spherical	40 nm	294 nm	Mansouri-Tehrani et al. (2021)
Sargassum bovinum	10 mL of DLE	50 mL of 1 mM PdCl$_2$	24 h	60 °C	–	160 rpm	PdNPs	Monodispersed and octahedral	5–10 nm	300 nm	Momeni and Nabipour (2015)
Sargassum cinereum							AgNPs	Triangular	45–76 nm	408 nm	Mohandass et al. (2013)
Sargassum cymosum	10^4 mg L^{-1} DLE	HAuCl$_4$	30 min – 4 h	21 °C	2–12	300–1000 rpm	AuNPs	Spherical	5–22 nm	515 to 535 nm	Costa et al. (2020)
Sargassum filipendula	25 mL of DLE	20 ml of Titanium (IV) butoxide	2 h	50 °C	–	Stirring	AgNPs	–	–	–	Alarif et al. (2023)
Bostrychia tenella	25 mL of DLE	20 ml of Titanium (IV) butoxide	2 h	50 °C	–	Stirring	AgNPs	–	100 nm	–	Alarif et al. (2023)

(continued)

Table 1.1 (continued)

Marine plant	*Plant part and concentration	Precursor concentration	Reaction time	Synthesis temperature	Reaction pH	Reaction mixture stirring	Synthesized nanoparticles	Shape	Size range	UV–VIS spectrum	Reference
Laurencia obtusa	25 mL of DLE	20 mL of Titanium (IV) butoxide	2 h	50 °C	–	Stirring	AgNPs	–	–	–	Alarif et al. (2023)
Halimeda tuna	25 mL of DLE	20 mL of Titanium (IV) butoxide	2 h	50 °C	–	Stirring	AgNPs	–	–	–	Alarif et al. (2023)
Sargassum dentifolium	50 mL of DLE	50 mL of 1 mM $AgNO_3$	1 h	45 °C	–	Stirring	AgNPs	Spherical	113 nm	420 nm	Saber et al. (2017)
Sargassum glaucescens	10 mL DLE	10^{-3} M $HAuCl_4$	4 h	RT	–	Stirring	AuNPs		3.65 ± 1.69 nm	538 nm	Ajdari et al. (2016)
Sargassum ilicifolium	100 mL of DLE	100 ul of 1 M $AgNO_3$	20 min	60 °C			AgNPs	Spherical	33–40 nm	414 nm	Kumar et al. (2012a)
Sargassum ilicifolium	10 mL of FLE	1 mM of $AgNO_3$	72 h	RT		120 rpm	AgNPs	Spherical, cubical and hexagonal	–	440–460 nm	Suparna Roy and Anantharaman (2018a, b, c, d)
Sargassum ilicifolium	100 mL of DLE	1 mL $PdCl_2$	5 days	RT			PdNPs	Spherical	60–80 nm	Gradual disappearance of 280 nm	Prasad and Padmesh (2014)
Sargassum ilicifolium							AgNPs		27.9 nm		Devi et al. (2022)
Sargassum ilicifolium							AuNPs		9.36 nm		Devi et al. (2022)
Sargassum ilicifolium	10 mL of DLE	90 mL of 0.05 M aluminum sulfate	1 h	25 °C		Stirring	Al_2O_3 NPs	Spherical	20 ± 2.1 nm	227 nm	Koopi and Buazar (2018)
Sargassum incisifolium	2 mg of DLE	500 µL of 0.1 M $AgNO_3$	18 h		–	–	AgNPs	Spherical	3.36–53.08 nm	410–440 nm	Mmola et al. (2016)
Sargassum incisifolium	2 mg of DLE	500 µL of 0.1 M $AgNO_3$	18 h		–	–	AuNPs	Spherical	2.17–268.67 nm	530 nm	Mmola et al. (2016)
Sargassum latifolium	DLE	1 mM $ZnSO_4$	2 h	RT	–	–	ZnNPs	Spherical, granular, and square aggregation	22.31–95.16 nm	280 nm	El-Khateeb et al. (2019)
Sargassum latifolium	DLE	1 mM SeO_2	2 h	RT	–	–	SeNPs	Spherical	22.31–95.16 nm	280 nm	El-Khateeb et al. (2019)
Sargassum longifolium							CuONPs	Spherical	40–60 nm	350 nm	Rajeshkumar et al. (2021)

(continued)

1.5 Green Synthesis of MFNPs

Table 1.1 (continued)

Marine plant	*Plant part and concentration	Precursor concentration	Reaction time	Synthesis temperature	Reaction pH	Reaction mixture stirring	Synthesized nanoparticles	Shape	Size range	UV-VIS spectrum	Reference
Sargassum longifolium							AgNPs	Cubical	30 nm	410 nm	Devi et al. (2013)
Sargassum longifolium	DLE	1 mM AgNO$_3$	–	RT	8.4	120 rpm	AgNPs	Spherical	40–85 nm	440 nm	Rajeshkumar et al. (2014)
Sargassum muticum	50 mL of DLE	50 mL of 1 mM AgNO$_3$	30 min	35 °C	7.5		AgNPs	Spherical	5–15 nm	420 nm	Azizi et al. (2013)
Sargassum muticum	50 mL of DLE	10 mL of 2 mM of zinc acetate di-hydrate	3–4 h	70 °C		Stirring	ZnONPs				Sanaeimehr et al. (2018)
Sargassum muticum	50 mL of DLE	2 mM Zn(Ac)$_2 \cdot$2H$_2$O	3–4 h	70 °C			ZnONPs	Hexagonal	30–57 nm	334 nm	Azizi et al. (2014)
Sargassum muticum	DLE	0.1 M FeCl$_3$	60 min	RT	–	Stirring	Fe$_3$O$_4$NPs	Cubic	18 ± 4 nm	402 nm and 415 nm	Mahdavi et al. (2013)
Sargassum muticum	FLE	0.1 M FeCl$_3$	60 min	25 °C		Stirring	Fe$_3$O$_4$MNPs	Cubic	18 ± 4 nm	402 nm and 415 nm	Namvar et al. (2015b)
Sargassum muticum	50 mL of FLE	50 mL of 0.1 M HAuCl$_4$	1 h	45 °C		Stirring	AuNPs	Spherical	5.42 ± 1.18 nm	520 nm	Namvar et al. (2015b)
Sargassum muticum	50 mL of Freeze DLE	50 mL of 0.1 M HAuCl$_4$	1 h	45 °C	–	Stirring	AuNPs	Spherical	5.42 ± 1.18 nm	550 nm	Namvar et al. (2015a)
Sargassum muticum	FLE	1 mM AgNO$_3$		RT			AgNPs	Spherical	43–79 nm	438 nm	Madhiyazhagan et al. (2015)
Sargassum muticum	5 mL of DLE	95 mL of 1 mM AgNO$_3$	20 min	90 °C			AgNPs	Spherical	42.30–98.56 nm	420 nm	Moorthi et al. (2015)
Sargassum muticum	0.5 gm/ mL FLE	0.25 mM AgNO$_3$	30 min	RT	4.81	Stirring	AgNPs	Spherical	10.4 ± 1.2 nm	427 nm	González-Ballesteros et al. (2020)
Sargassum muticum	1 gm/ mL FLE	0.4 M HAuCl$_4$	24 h	RT	4.38	Stirring	AuNPs	Spherical	41.0 ± 5.7 nm	538 nm	González-Ballesteros et al. (2020)
Sargassum muticum	100 mL of DLE	1 mM AgNO$_3$	24 h	RT	–	–	AgNPs	Spherical	20–54 nm	400 nm	Trivedi et al. (2021)
Sargassum myriocystum	5 mL of DLE	95 mL of 1 mM zinc nitrate	1 h – 6 months	50–100 °C	5–10	Magnetic stirred	ZnONPs	Triangle, radial hexagonal, rod, and rectangle	36–186 nm	372 nm	Nagarajan and Kuppusamy. (2013)
Sargassum myriocystum	2.5 g of DLE	100 mL of 10^{-3} M HAuCl$_4$					AuNPs	Cubical	40–85 nm	550 nm	Ismail et al. (2018)

(continued)

Table 1.1 (continued)

Marine plant	*Plant part and concentration	Precursor concentration	Reaction time	Synthesis temperature	Reaction pH	Reaction mixture stirring	Synthesized nanoparticles	Shape	Size range	UV–VIS spectrum	Reference
Sargassum myriocystum	10 mL of DLE	90 mL of 1 M $AgNO_3$	30 min–32 h	30–50° C	6.2–9.2	Magnetic stirrer	AgNPs	Spherical	25–65 nm	430 nm	Kumar and Rajeshkumar (2017)
Sargassum myriocystum	100 mL of DLE	900 mL of 1 mM $AgNO_3$	24 h	–	–	–	AgNPs	Hexagonal	20 ± 2.2 nm	420 nm	Balaraman et al. (2020)
Sargassum myriocystum	15 mL of DLE	85 mL of 5 mM titanium dioxide	–	–	–	–	TiO_2NPs	Cubic, square, and spherical	50–90 nm	403 nm	Balaraman et al. (2022)
Sargassum myriocystum	600 μL of DLE	10 mL of 1 mM $HAuCl_4$	15 min	72 °C	–	Water bath	AuNPs	Polydispersed triangular and spherical	15 nm	533 nm	Dhas et al. (2012)
Sargassum plagiophyllum	1 mL of DLE	99 mL of 1 mM $AgNO_3$	12 h	RT	–	–	AgNPs	Spherical	20 nm	420 nm	Dhanalakshmi et al. (2012)
Sargassum plagiophyllum	50 to 500 μL of DLE	5 mL of 1 mM $AgNO_3$	24 h	RT	–	–	AgClNPs	Spherical	18–48 nm	417 nm	Dhas et al. (2014a, b)
Sargassum polycystum	5 mL of DLE	95 mL of 10^{-3} M $HAuCl_4$	30 min	–	–	150 rpm	AuNPs	Spherical	30–60 nm	532 nm	Sivaraj et al. (2015)
Sargassum polycystum	5 mL of DLE	95 mL of 1 mM $AgNO_3$	–	20 °C	–	–	AgNPs	Spherical	~28 nm	405 nm	Palanisamy et al. (2017)
Sargassum polycystum	0.3 gm of DLE in 50–60 mL of deionized water	0.2, 0.4, 0.6 and 0.8 mL of 1 M $AgNO_3$	24 h	RT	–	–	AgNPs	Spherical and elongated	5–7 nm	420 nm	Thangaraju et al. (2012)
Sargassum polycystum	50 gm DLE	90 mL of 1 mM $AgNO_3$	–	60 °C	–	Magnetic stirrer	AgNPs	Cubical	20–88 nm	418 nm	Vinoth et al. (2019)
Sargassum polycystum	DLE	100 mL of 1 mM copper solution	Overnight	RT	–	–	CuONPs				Ramaswamy et al. (2016)
Sargassum polycystum	10 gm FLE	1 mM $AgNO_3$	–	–	–	–	AgNPs	–	–	430 nm	Asha et al. (2015)
Sargassum polycystum	10 mL of DLE	90 mL of 0.1 mM $AgNO_3$	24–48 h	RT	–	–	AgNPs	Cluster and spherical	>100 nm	422 nm	Thirunavukkarau et al. (2022)

(continued)

1.5 Green Synthesis of MFNPs

Table 1.1 (continued)

Marine plant	*Plant part and concentration	Precursor concentration	Reaction time	Synthesis temperature	Reaction pH	Reaction mixture stirring	Synthesized nanoparticles	Shape	Size range	UV–VIS spectrum	Reference
Sargassum polyphyllum	10 mL of FLE	90 mL of 1 mM AgNO$_3$	24 h	25–37 °C		100 rpm	AgNPs	Spherical	37–43 nm	420 nm	Arunkumar et al. (2014)
Sargassum siliquosum	2–10% w/v of DLE	6 mM AgNO$_3$	30 min	Varied			AgNPs	Spherical	20–480 nm	420–460 nm	Vasquez et al. (2016)
Sargassum swartzii	1 mL of WE	5 mL of 1 mM HAuCl$_4$		60 °C			AuNPs	Spherical	35 nm	527 nm	Stalin Dhas et al. (2014a, b)
Sargassum swartzii	80 mL of DLE	400 mL of 0.1 M Zn(CH$_3$COO)$_2$·2H$_2$O	3 h	70 °C		Magnetic stirring	ZnONPs	Spherical	32 nm	395 nm	Vinu et al. (2021)
Sargassum swartzii	40 mL of DLE	0.1 M CuCH$_3$COO)$_2$·H$_2$O	4 h	RT		Magnetic stirring	CuONPs	Spherical	32 nm	260 nm	Vinu et al. (2021)
Sargassum swartzii	100 mL of DLE	3 mM H$_2$SeO$_3$ + 150 mL of ascorbic acid				Magnetic stirring	SeNPs	Spherical	21 nm	233–258 nm	Vinu et al. (2021)
Sargassum swartzii	1 mL of DLE	5 mL of 10^{-3} M AgNO$_3$	24 h	30 °C	–	–	AgNPs	Spherical	20–40 nm	404 nm	Dhas et al. (2021)
Sargassum tenerrimum	5 mL DLE	45 mL of 1 mM HAuCl$_4$		RT		Magnetic stirrer	AuNPs	Spherical	5–45 nm	547 nm	Ramakrishna et al. (2016)
Sargassum tenerrimum	5 mL of DLE	95 mL of 1 mM AgNO$_3$	20 min	90 °C			AgNPs	Spherical	20 nm	420 nm	Kumar et al. (2012a, b, c)
Sargassum vulgare	Alginate from DLE	10 mL of 10^{-3} M AgNO$_3$	3 h	RT			AgNPs	Spherical	10 nm	419 nm	Govindaraju et al. (2015)
Sargassum vulgare	DLE	0.1 M FeCl$_3$	1 h 30 min	RT	–	800 rpm	Fe$_3$O$_4$NPs	Nanospheres	17.05–34.09 nm	402 and 415 nm	Salem et al. (2020)
Sargassum wightii							AgNPs	Spherical	18.45–41.59 nm	420 nm	Deepak et al. (2018)
Sargassum wightii	5 mL of DLE	50 mL of AgNO$_3$	24 h	28 °C	–	–	AgNPs	Spherical	5–22 nm	439 nm	Shanmugam et al. (2014)
Sargassum wightii	10 mL of DLE	90 mL of 10^{-3} M AgNO$_3$	–	RT	–	–	AgNPs	Spherical	8–27 nm	431 nm	Govindaraju et al. (2009)

(continued)

Table 1.1 (continued)

Marine plant	*Plant part and concentration	Precursor concentration	Reaction time	Synthesis temperature	Reaction pH	Reaction mixture stirring	Synthesized nanoparticles	Shape	Size range	UV–VIS spectrum	Reference
Sargassum wightii	1 gm of DLE	100 mL of 10^{-3} M HAuCl$_4$	12 h	–	–	Stirring	AuNPs	Planar structures	8–12 nm	527 nm	Singaravelu et al. (2007)
Sargassum wightii	1 gm of DLE	100 mL of 10^{-3} M HAuCl$_4$	48 h	30 °C	6	Vigorous stirring	AuNPs	Star shaped	30–100 nm	585 nm	Oza et al. (2012)
Sargassum wightii	DLE	100 mL of 10^{-3} M AgNO$_3$	10 min	45 °C	–	Magnetic stirrer	AgNPs	Circular	40 nm	438 nm	Sunitha et al. (2015)
Sargassum wightii	DLE	10^3 M AgNO$_3$					AgNPs		55–70 nm	407 nm	Selvaraj et al. (2020)
Sargassum wightii	FLE	Mg(NO$_3$)$_2$	6 h	90 °C		Stirring	MgONPs	Flower	68.06 nm	322 nm	Pugazhendhi et al. (2019)
Sargassum wightii	DLE	90 mL of 1 mM AgNO$_3$	3 h	RT	–	Vigorous stirring	AgNPs	Spherical	80–100 nm	420 nm	Suganya et al. (2020)
Sargassum wightii	10 mL of DLE	90 mL of 1 mM AgNO$_3$	5 h	RT	–	–	AgNPs	Spherical and cubic	25–50 nm	~465 nm	Selvam and Sivakumar (2014)
Sargassum wightii	10 mL of DLE	90 mL of 2 mM AgNO$_3$	2 h	RT	–	Stirring	AgNPs	Spherical	48.78 nm	450 nm	Thirumalairaj et al. (2014)
Sargassum wightii	5 mL of DLE	95 mL of 1 mM zinc acetate	3–4 h	70 °C	–	Magnetic stirring	ZnONPs	Spherical	40–50 nm	327 nm	Ishwarya et al. (2018b)
Sargassum wightii	FLE	0.05 mM zinc nitrate	4–5 h	90 °C	–	Stirring	ZnONPs	Spherical	20–62 nm	378 nm	Murugan et al. (2018)
Sargassum wightii	1 gm of DLE	5 mM of ZrO(NO$_3$)$_2$·xH$_2$O	20 min	400 ± 10 °C	–	Grounded	ZrO$_2$-NPs	Spherical	5 nm	277 nm	Kumaresan et al. (2018)
Sargassum polycystum	10 mL of DLE	90 mL of 0.1 mM AgNO$_3$	24–48 h	RT	–	–	AgNPs	–	–	411 nm	Thiurunavukkarau et al. (2022)
Scaberia agardhit	DLE	1 mM AgNO$_3$			–	–	AgNPs	Polydispersed	40–50 nm	409 nm	Prasad and Elumalai (2013)
Spatoglossum asperum	DLE	1 mM AgNO$_3$					AgNPs	Spherical	10–50 nm	420 nm	Subbiah et al. (2019)

(continued)

1.5 Green Synthesis of MFNPs

Table 1.1 (continued)

Marine plant	*Plant part and concentration	Precursor concentration	Reaction time	Synthesis temperature	Reaction pH	Reaction mixture stirring	Synthesized nanoparticles	Shape	Size range	UV–VIS spectrum	Reference
Spatoglossum asperum	10 mL of fucoidan (1 to 10 mg/mL)	90 mL of 1 mM AgNO$_3$	12 h	37 °C	–	Static	AgNPs	Spherical to oval	20–46 nm	440 nm	Ravichandran et al. (2018)
Spatoglossum schroederi	10 mg/mL fucan A solution	0.001 M AgNO$_3$	1 h	–	–	–	AgNPs	Spherical	210 nm	–	Amorim et al. (2016)
Staechospermum marginatum	1 gm of DLE	100 mL of 10^{-3} M HAuCl$_4$	12 s	–	–	Stirring	AuNPs	Spherical, hexagonal and triangle	18.7–93.7 nm	550 nm	Rajathi et al. (2012)
Turbinaria conoides	DLE	100 mg/L of HAuCl$_4$				200 rpm	AuNPs	Sporadic aggregation	20–80 nm	540 nm	Vijayaraghavan et al. (2011)
Turbinaria conoides	10 mL of DLE	100 mL of 1 mM HAuCl$_4$	8 h	37 °C	–	HM	AuNPs	Spherical and cubic	< 1 μm	599.78 nm	Venkatraman et al. (2018)
Turbinaria conoides	5 mL DLE	45 mL of 1 mM HAuCl$_4$		RT		Magnetic stirrer	AuNPs	Spherical	12–57 nm	536 nm	Ramakrishna et al. (2016)
Turbinaria conoides	10 mL of DLE	90 mL of 1 mM AgNO$_3$		RT		HM	AgNPs	Spherical	96 nm	420 nm	Rajeshkumar et al. (2012b)
Turbinaria conoides	10 mL of DLE	90 mL of 1 mM gold chloride	10 min				AuNPs	Spherical-triangle	6–10 nm	520 to 525 nm	Shanmugam Rajeshkumar et al. (2013a, b)
Turbinaria conoides	10 mL of FLE	100 mL of 1 mM HAuCl$_4$	50 min – 48 h	RT			AuNPs	Square, rectangle, cubic and triangle	60 nm		Rajeshkumar et al. (2013a, b)
Turbinaria conoides	DLE	90 mL of 1 mM AgNO$_3$		RT			AgNPs	Spherical	2–17 nm	421 nm	Vijayan et al. (2014)
Turbinaria conoides	DLE	90 mL of 1 mM HAuCl$_4$		RT			AuNPs	Triangular	2–19 nm	538 nm	Vijayan et al. (2014)
Turbinaria conoides							ZnONPs				Rajeshkumar (2018)
Turbinaria conoides							ZnONPs				Raajshree and Brindha (2018)

(continued)

Table 1.1 (continued)

Marine plant	*Plant part and concentration	Precursor concentration	Reaction time	Synthesis temperature	Reaction pH	Reaction mixture stirring	Synthesized nanoparticles	Shape	Size range	UV–VIS spectrum	Reference
Turbinaria decurrens	10 mL of DLE	1.99 g of $FeCl_2 \cdot 4H_2O$ and 5.41 g $FeCl_3 \cdot 6H_2O$	–	–	10	–	SPIONPs	Spherical	18 nm	–	Khaleefullah et al. (2017)
Turbinaria ornata	FLE	0.1, 0.25, 0.5, 0.75, 1 mM of $AgNO_3$	48 h	37 °C	–	–	AgNPs	Cubical	18–20 nm	420 nm	Kayalvizhi et al. (2014)
Turbinaria ornata	FLE	0.1, 0.25, 0.5, 0.75, 1 mM of $HAuCl_4$	48 h	37 °C	–	–	AuNPs	Cubical	20–90 nm	550 nm	Kayalvizhi et al. (2014)
Turbinaria ornata	5 mL of DLE	95 mL of 1 mM $AgNO_3$		70 °C	–		AgNPs	Cubic and hexagonal	49–75 nm	420 nm	Krishnan et al. (2015)
Turbinaria ornata	12 mL of DLE	88 mL of 1 mM $AgNO_3$		RT	–		AgNPs	Spherical	20–32 nm	420 nm	Deepak et al. (2017)
Turbinaria ornata	10 mL of DLE	90 mL of 1 mM $AgNO_3$	–	RT	–	120 rpm	AgNPs	Spherical	14–22 nm	428 nm	Remya et al. (2017)
Turbinaria ornata	(Deepak et al. 2017; Abdel Azeem et al. 2021)	Refer					AgNPs	Spherical	20–60 nm	436 nm	Azeem et al. (2022)
Turbinaria ornata	10 mL of DLE	90 mL of 5 mM $AgNO_3$	24 h	70 °C	11	Stirring	AgNPs	Spherical	30–40 nm	430 nm	Anuluxan et al. (2022)
Turbinaria turbinata	10 mL of DLE	90 mL of 1 mM $AgNO_3$	40 min	30 °C	–	Magnetic stirring	AgNPs	Spherical	8–16 nm	428 nm	Bialy et al. (2017)
Turbinaria turbinata	10 mL of DLE	90 mL of 1 mM $AgNO_3$	40 min	30 °C	–	Magnetic stirring	AgNPs	Spherical	8–16 nm	428 nm	Khalifa et al. (2016)
Undaria pinnatifida	4 mL of fucoidan solution	0.8 mL of lactoferrin solution	30 min	–	–	400 rpm	Fucoidan/Lactoferrin (FCD/LF) NPs	Spherical	$167 \pm 4.2 - 346.3 \pm 12.6$ nm	–	Etman et al. (2020)

(continued)

1.5 Green Synthesis of MFNPs

Table 1.1 (continued)

Marine plant	*Plant part and concentration	Precursor concentration	Reaction time	Synthesis temperature	Reaction pH	Reaction mixture stirring	Synthesized nanoparticles	Shape	Size range	UV–VIS spectrum	Reference
Undaria pinnatifida	Fucoidan solution	4 mg/mL solution of Quinacrine	30 min	–	–	Stirring	Fucoidan/ quinacrine (FCD/QC) NPs	Spherical	> 200 nm	425 nm	Etman et al. (2021)
Undaria pinnatifida	1 mL of 1% polysaccharide solution + 8 mL 60 mM ascorbic acid solution	1 mL of 30 mM sodium selenite solution	5 min	–	–	Sonication	SeNPs	Spherical	44–92 nm	–	Chen et al. (2008)
Red seaweeds											
Marine red algae	0.11 gm of Carrageenan Oligosaccharide	10 mL of 6 × 10^{-4} mol/L $HAuCl_4 \cdot 3H_2O$	3 h	50 °C	–	Stirring	AuNPs	Ellipsoidal	35 ± 8 nm	530 nm	Chen et al. (2018)
Marine red algae	10 mL of DLE	45 mL of 1 mM $Co(NO_3)_2 \cdot 6H_2O$	24 h	RT	–	Stirring	Co_3O_4 NPs	Spherical	29.8 ± 8.6 nm	508 nm	Ajarem et al. (2022)
Acanthophora spicifera	DLE	1 mM $AgNO_3$	20 min	60 °C	–	–	AgNPs	Spherical	48 nm	437 nm	Kumar et al. (2012b)
Acanthophora spicifera							AgNPs	Cubical	33–81 nm	–	Ibraheem et al. (2016)
Acanthophora spicifera	250 mL of DLE	65 μL of 1 M $HAuCl_4$	4 h	60 °C	–	Stirring	AuNPs	Spherical-to-oval	< 20 nm	536 nm	Babu et al. (2020)
Acanthophora spicifera	10 mL of DLE	90 mL of 0.1 mM $AgNO_3$	24–48 h	RT	–	–	AgNPs	–	–	429 nm	Thiurunavukkarau et al. (2022)
Actinotrichia fragilis	9.5 mL of DLE	50–500 μL of 0.05 M $HAuCl_4$	10 h	RT	–	500 rpm	AuNPs	Triangle, truncated triangular, hexagon and polygon	10–20 nm	500 nm	Momeni et al. (2016)
Amphiroa anceps	10 mL of FLE	1 mM of $AgNO_3$	72 h	RT	–	120 rpm	AgNPs	Spherical	10–80 nm	420 nm	Roy and Anantharaman (2018c)

(continued)

Table 1.1 (continued)

Marine plant	*Plant part and concentration	Precursor concentration	Reaction time	Synthesis temperature	Reaction pH	Reaction mixture stirring	Synthesized nanoparticles	Shape	Size range	UV–VIS spectrum	Reference
Amphiroa fragilissima	DLE	1 mM $AgNO_3$	–	–	–	–	AgNPs	Triangular, hexagonal and spherical	5–50 nm	429.5 nm	Ramalingam et al. (2018)
Amphiroa rigida							AgNPs				Gopu et al. (2021)
Centroceras clavulatum	DLE	1 mM $AgNO_3$	–	RT	–	–	AgNPs	Spherical	35–65 nm	410 nm	Murugan et al. (2016a)
Champia parvula	10 mL of DLE	90 mL of 1 mM $AgNO_3$	1 h	37 °C	–	–	AgNPs	Round-shaped	79 nm	425 nm	Viswanathan et al. (2023)
Chondrus crispus	0.25 gm of DLE	50 mL of 10^{-3} M $HAuCl_4$	–	RT	10	Stirring	AuNPs	Spherical, triangular, hexagonal, and wire like	30–200 nm	540 nm	Castro et al. (2013)
Chondrus crispus	0.25 gm of DLE	50 mL of 10^{-3} M $AgNO_3$	–	RT	10	Stirring	AgNPs	Spherical	30 nm	460 nm	Castro et al. (2013)
Chondrus crispus	1 g/mL DLE	0.5 mM $HAuCl_4$	48 h	30 °C	–	–	AuNPs	Spherical	10–26 nm	529 nm	González-Ballesteros et al. (2022)
Chondrus crispus	Carrageenan	0.01 mM $HAuCl_4$	24 h	90 °C	11	Stirring	AuNPs	Decahedral quasi-spherical	8.4 nm	525 nm	Alvarez-Vinas et al. (2022)
Corallina elongata	10 mL of DLE	90 mL of 0.017 gm of $AgNO_3$	30 min	60 °C	–	Magnetic stirring	AgNPs	Spherical	8–25 nm	407–410	Hamouda et al. (2019)
Corallina elongata	10 mL of DLE	90 mL of 1 mM $AgNO_3$	40 min	30 °C	–	Magnetic stirring	AgNPs	–	–	428 nm	Khalifa et al. (2016)
Corallina mediterranea							AgNPs	Spherical	4–80 nm	430 nm	Negm et al. (2018)
Corallina officinalis							AuNPs	Spherical	14.6 ± 1 nm	530 nm	El-Kassas and El-Sheekh (2014)
Galaxaura elongata	1 gm of DLE	100 mL of 10^{-3} M $HAuCl_4$		RT			AuNPs	Spherical; few rod, triangular, truncated triangular and hexagonal	3.85–77.13 nm	535 nm	Abdel-Raouf et al. (2017)

(continued)

1.5 Green Synthesis of MFNPs

Table 1.1 (continued)

Marine plant	*Plant part and concentration	Precursor concentration	Reaction time	Synthesis temperature	Reaction pH	Reaction mixture stirring	Synthesized nanoparticles	Shape	Size range	UV-VIS spectrum	Reference
Galaxaura elongata	50 mL of DLE	450 mL of 10^{-3} M AgNO$_3$	–	–	–	–	AgNPs	Spherical	30–90 nm	440 nm	Azeem et al. (2022)
Gelidiella sp.	900 mL of FLE	100 mL of 1 mM AgNO$_3$	10 min	121 °C	–	Magnetic stirring	AgNPs	Spherical	40–50 nm	435 nm	Devi et al. (2012)
Gelidiella acerosa							AuNPs	Spherical, hexagonal, & crystalline	5.81–117.59 nm	526 nm	Senthilkumar et al. (2019)
Gelidiella acerosa							AuNPs	Spherical	5–20 nm		Subbulakshmi et al. (2022)
Gelidiella acerosa							AgNPs	Spherical	22 nm	408 nm	Vivek et al. (2011)
Gelidiella acerosa	10 mL of DLE	100 mL of 1 mM AgNO$_3$	6 h	37 °C	–	180 rpm	AgNPs	Spherical	20–50 nm	418 nm	Satish et al. (2017)
Gelidiella acerosa	20 mL of DLE	80 mL of 1 mM AgNO$_3$	24 h	RT	–	–	AgNPs	–	59.72 nm	404 nm	Thiruchelvi et al. (2021)
Gelidium amansii	2 mg of DLE	1 mM, 3 mM and 5 mM AgNO$_3$		Below 40° C	–	150 rpm	AgNPs	Spherical	27–54 nm	420 nm	Pugazhendhi et al. (2018)
Gelidium amansii	10 mL of DLE	90 mL of 0.017 gm of AgNO$_3$	30 min	60 °C	–	Magnetic stirring	AgNPs	Square	12–20 nm	413–410 nm	Hamouda et al. (2019)
Gelidium amansii	2 mg of DLE	1 mL of 0.5 mM of HAuCl$_4$	1 h	RT	–	–	AuNPs	Cubical	4–7 nm	540 nm	Kumar et al. (2017a, b)
Gelidium corneum	DLE	AgNO3					AgNPs	Spherical	20–50 nm	420–430 nm	Öztürk et al. (2020)
Gelidium corneum	1 g/mL DLE	0.4 mM HAuCl$_4$	24 h	30 °C	–	–	AuNPs	Polyhedral	30–58 nm	529 nm	González-Ballesteros et al. (2022)
Gelidium crinale	10 mL of DLE	90 mL of 1 mM AgNO$_3$	40 min	30 °C	–	Magnetic stirring	AgNPs	–	–	428 nm	Khalifa et al. (2016)
Gelidium pusillum	25 mL of DLE	1 mM HAuCl$_4$	20 min	60 °C	–	Stirring	AuNPs	Spherical	12 ± 4.2 nm	529 nm	Jeyarani et al. (2020)

(continued)

Table 1.1 (continued)

Marine plant	*Plant part and concentration	Precursor concentration	Reaction time	Synthesis temperature	Reaction pH	Reaction mixture stirring	Synthesized nanoparticles	Shape	Size range	UV–VIS spectrum	Reference
Gracilaria sp.	50 mL of DLE	10^{-3} M AgNO3	120 h		5.0	100 rpm	AgNPs	Round and spherical	10–40 nm	419 nm	Ramakritinan et al. (2013)
Gracilaria sp.	50 mL of DLE	10^{-3} M HAuCl4	120 h		5.0	100 rpm	AuNPs	Round, spherical and disbursed	10–30 nm	536 nm	Ramakritinan et al. (2013)
Gracilaria birdiae							AgNPs	Spherical	20.2–94.9 nm	410 nm	Aragao et al. (2019)
Gracilaria corticata	30 gm of DLE	10 mL of 10^{-3} M HAuCl4		RT			AuNPs	Hexagonal and spherical	5–135 nm	550 nm	Sugandhi and Rani (2014)
Gracilaria corticata	10 mL of DLE	90 μL of 10^{-3} M HAuCl4	4 h	40 °C		120 rpm	AuNPs		45–57 nm	540 nm	Naveena and Prakash (2013)
Gracilaria corticata	DLE	1 mM AgNO3	10 min	121 °C	–	–	AgNPs	Spherical	10–50 nm	424 nm	Bhimba et al. (2015)
Gracilaria corticata	10 mL of DLE	90 μL of 1 M AgNO3	24 h	Varied			AgNPs	Spherical	100 nm	380 nm	Poornima and Valivittan (2017)
Gracilaria corticata							AgNPs	Spherical	37 nm	420 nm	Roseline et al. (2019)
Gracilaria corticata							AgNPs	Spherical	20 ± 5 nm	389 nm	Parthasarathy et al. (2021)
Gracilaria corticata	99.9 mL of DLE	100 μL of 1 M AgNO3	20 min	60 °C			AgNPs	Spherical	18–46 nm	420 nm	(Kumar et al. 2013b)
Gracilaria corticata	10 mL of DLE	50 mL of AgNO3	24 h	60 °C	–	160 rpm	AgNPs	Spherical and monodispersed	24.18 nm	427 nm	Naveenkumar et al. (2023)
Gracilaria corticata	5 mL of DLE	95 mL of 10^{-3} M AgNO3	1 h	100 °C	–	–	AgNPs	Spherical	–	420 nm	Aravindan et al. (2014)
Gracilaria crassa	5 mL of DLE	95 mL of 1 mM AgNO3	Overnight	30 °C	–	Stirring	AgNPs	Spherical	122.7 nm	443 nm	Lavakumar et al. (2015)
Gracilaria crassa	30 mL DLE	70 mL of 1 mM HAuCl4	Few minutes	RT	7.2	Shaker	AuNPs	Spherical	32.0 nm ± 4.0 nm	547(Shukla et al. 2012) nm	Kamaraj et al. (2022)

(continued)

1.5 Green Synthesis of MFNPs

Table 1.1 (continued)

Marine plant	*Plant part and concentration	Precursor concentration	Reaction time	Synthesis temperature	Reaction pH	Reaction mixture stirring	Synthesized nanoparticles	Shape	Size range	UV–VIS spectrum	Reference
Gracilaria dura	1, 3 and 5 mg mL^{-1} agar powder solution	2.5 and 5 mM AgNO$_3$	1, 4 and 48 h	25, 60, and 100 °C	6	Stirring	AgNPs	Spherical	6 nm	421 nm	Shukla et al. (2012)
Gracilaria dura	5 mL of DLE	25 mL of 3 mM AgNO$_3$	24 h	RT	–	–	AgNPs	Spherical	–	446 nm	Paul and Devi (2014)
Gracilaria edulis	500 mg of DLE	10^{-3} M AgNO$_3$	24 h	37 °C	5.6	Stirring	AgNPs	Spherical	12.5–100 nm	403 nm	Murugesan et al. (2011)
Gracilaria edulis							AgNPs	Spherical	54 nm	430 nm	Roseline et al. (2019)
Gracilaria edulis							AgNPs	Spherical	55–99 nm	430 nm	Priyadharshini et al. (2014)
Gracilaria edulis							ZnONPs	Rod	66–95 nm	335 nm	Priyadharshini et al. (2014)
Gracilaria edulis	DLE	1 mM AgNO$_3$	10 min – 4 h	95 °C	–	–	AgNPs	Spherical and cubic	30–42 nm	342–412 nm	Madhiyazhagan et al. (2017)
Gracilaria edulis	10 mL of FLE	90 mL of 10 mM Zn(NO$_3$)$_2$·6H$_2$O	5 h	60 °C	–	Stirring	ZnONPs	Wurtzite structure (hexagonal phase)	20–50 nm	367 nm	Asik et al. (2019)
Gracilaria firma	DLE	1 mM AgNO$_3$		RT	–		AgNPs	Spherical	12–200 nm	440 nm	Kalimuthu et al. (2017)
Gracilaria foliifera	WE	10^{-3} M	–	RT	–	HM	AuNPs	Spherical	13 nm	543 nm	Algotiml et al. (2022a)
Gracilaria foliifera	20 mL of DLE	80 mL of 10^{-3} M AgNO$_3$	15 min	70 °C	–	Stirred on a hot plate	AgNPs	Spherical	24 nm	415 nm	Algotiml et al. (2022b)
Gracilaria gracilis	DLE	AgNO$_3$					AgNPs		12–46 nm		Kochesfehani et al. (2021)
Gracilaria gracilis	0.254 g Agar	0.127 g Zn(NO$_3$)$_2$	30 min	–	–	350 rpm in ball mill	ZnONPs	Hexagonal-like	50 nm average	–	Francavilla et al. (2014)
Gracilaria gracilis	8 g Alginic acid	4 g Zn(NO$_3$)$_2$	30 min	–	–	350 rpm in ball mill	ZnONPs	Flower-like	20 nm average	–	Francavilla et al. (2014)
Gracilaria gracilis	6 g Starch	3 g Zn(NO$_3$)$_2$	15 min	–	–	650 rpm in ball mill	ZnONPs	Flower-like and hexagonal rod	26 nm average	–	Francavilla et al. (2014)

(continued)

Table 1.1 (continued)

Marine plant	*Plant part and concentration	Precursor concentration	Reaction time	Synthesis temperature	Reaction pH	Reaction mixture stirring	Synthesized nanoparticles	Shape	Size range	UV–VIS spectrum	Reference
Gracilaria parvispora	1 gm of DLE	100 mL of 10^{-3} M $AgNO_3$	12 h	–	–	Stirring	AgNPs	Spherical	12–30 nm	412 nm	Hussein et al. (2017)
Grateloupia sp.							AgNPs	Spherical	4–80 nm	430 nm	Negm et al. (2018)
Halymenia porphyroides	500 mg of DLE	10^{-3} M $AgNO_3$	24 h	–	5.09		AgNPs	Spherical	34–80 nm	420 nm	Kiran and Murugesan (2014b)
Halymenia porphyroides	–	–	–	–	–	–	AgNPs	–	–	–	Vishnu and Murugesan (2014)
Halymenia porphyroides	500 mg of DLE	10^{-3} M $AgNO_3$	24 h	RT	5.09	Stirring	AgNPs	Spherical	34.3–80 nm	430.5 nm	Manam and Subbaiah (2020)
Halymenia porphyriformis	DLE	0.01 M $AgNO_3$	24 h	RT	–	–	AgNPs	Cubic and spherical	15.23 nm	420 nm	Khan et al. (2022a, b)
Halymenia pseudofloresii	30 mL of DLE	70 mL of 1 mM $HAuCl_4$	20 min	60 °C	–	Stirring	AuNPs	Cubic and rectangular	27 nm	545 nm	Palaniyandi et al. (2023)
Halymenia venusta	30 mL of DLE	70 mL of 1 mM $HAuCl_4$	2 h	60 °C	–	Stirring	AuNPs	Spherical	81 nm	510 nm	Baskar et al. (2023)
Hypnea musciformis							AgNPs	Spherical	53 nm	410 nm	Roseline et al. (2019)
Hypnea musciformis	FLE	1 mM $AgNO_3$					AgNPs	Spherical	40–65 nm	420 nm	Roni et al. (2015)
Hypnea musciformis	1 gm of DLE	1 mM $AgNO_3$				Agitated	AgNPs	Spherical		429 nm	Devi and Bhimba (2014)
Hypnea musciformis	500 mg of DLE	10^{-3} M $HAuCl_4$		RT	5.09		AuNPs	Spherical	6.25–33.33 nm	552 nm	Murugesan et al. (2015)

(continued)

1.5 Green Synthesis of MFNPs

Table 1.1 (continued)

Marine plant	*Plant part and concentration	Precursor concentration	Reaction time	Synthesis temperature	Reaction pH	Reaction mixture stirring	Synthesized nanoparticles	Shape	Size range	UV–VIS spectrum	Reference
Hypnea musciformis	5 mL of DLE	50 mL of AgNO$_3$	24 h	28 °C	–	–	AgNPs	Cubical	2–55.8 nm	440 nm	Selvam and Sivakumar (2015)
Hypnea musciformis	10 mL of DLE	190 mL of 1 mM AgNO3	3 h	–	–	Sunlight irradiation	AgNPs	Triangles, pentagons and hexagons and spheres	16–42 nm	420 nm	Vadlapudi and Amanchy (2017)
Hypnea valentiae	5 mL of DLE	95 mL of 1 mM zinc nitrate	1 h – 6 months	50–100 °C	5–10	Stirring	ZnONPs	Triangle, radial, hexagonal, rod, and rectangle	36–186 nm	372 nm	Nagarajan and Kuppusamy (2013)
Hypnea valentiae	10 mL of DLE	90 mL of 1 mM AgNO$_3$	1 h	RT	–	–	AgNPs	Spherical	10–45 nm	430 nm	Viswanathan et al. (2022)
Iridaea cordata	0.13 gm/mL of FLE	0.5 mM HAuCl$_4$	40 s	RT	3.87	Stirring	AuNPs	Spherical	12.3 ± 1.6 nm	539 nm	González-Ballesteros et al. (2021)
Iridaea cordata	0.2 gm/mL of FLE	0.17 mM AgNO3	1 h	100° C	6.17	Stirring	AgNPs	Spherical	17.5 ± 3.7 nm	418 nm	González-Ballesteros et al. (2021)
Jania rubins	90 mL of crude hot water soluble polysaccharide solution	1 mL of 0.1 mM AgNO$_3$	20 min	70 °C	10	Stirring	AgNPs	Spherical	12 nm	407–424 nm	El-Rafie et al. (2013)
Jania rubens	DLE	0.1 M FeCl$_3$	3 h 30 min	RT	–	800 rpm	Fe$_3$O$_4$NPs	Nanospheres	22.22–33.33 nm	402 and 415 nm	Salem et al. (2020)
Jania rubens	50 mL of DLE	50 mL of 1 mM AgNO$_3$	1 h	45 °C	–	Stirring	AgNPs	Spherical	155 nm	440 nm	Saber et al. (2017)
Kappaphycus	5 mL of CAO-AuNPs	3 mL of 0.8 mg/mL epirubicin	24 h	40 °C	–	–	epirubicin-loaded kappa-carrageenan gold (EPI-CAO-AuNPs) NPs	Spherical	141 ± 6 nm	490 nm	Chen et al. (2019)
Kappaphycus	5 mL of Kappa-Carrageenan + 50 mL of NaOH	50 mL of 2 M Zinc acetate	2 h	–	–	Stirring	ZnONPs	Spherical and hexagonal	97.03 ± 9.05	373 nm	Vijayakumar et al. (2020)

(continued)

Table 1.1 (continued)

Marine plant	*Plant part and concentration	Precursor concentration	Reaction time	Synthesis temperature	Reaction pH	Reaction mixture stirring	Synthesized nanoparticles	Shape	Size range	UV–VIS spectrum	Reference
Kappaphycus alvarezii	100 mg of DLE	100 mL of 10^{-3} M $HAuCl_4$	2 h	RT	–	Stirring	AuNPs	Spherical	10–40 nm	539 nm	Rajasulochana et al. (2010)
Kappaphycus alvarezii	100 mg of DLE	100 mL of 10^{-3} M $HAuCl_4$	2 h	–	–	Stirring	AuNPs	Oval, spherical, and irregular	10–40 nm	–	Rajasulochana et al. (2012)
Kappaphycus alvarezii	DLE	$CuSO_4 \cdot 5H_2O$	20 min	120 °C	–	Stirring	Cu@Cu_2ONPs	Spherical	52.99 ± 18.64 nm	590 nm and 390 nm	Khanehzaei et al. (2015)
Kappaphycus alvarezii	10 mL of DLE	190 mL of 1 mM $AgNO_3$	96 h	27 °C	–	250 rpm	AgNPs	Spherical	73 nm	420 nm	Ganesan et al. (2013)
Kappaphycus alvarezii	400 mL of FLE	100 mL of 10 mM $AgNO_3$	10–720 min	–	4.35–4.84	Ultrasound irradiation	AgNPs	Spherical	11.78 nm	384 nm	Faried et al. (2016)
Kappaphycus alvarezii	DLE	Fe^{3+} and Fe^{2+} with 2:1 M ratio	1 h	–	11	Stirring	Fe_3O_4 NPs	Spherical	14.7 nm	300–800 nm	Yew et al. (2016)
Kappaphycus alvarezii	10 mL of DLE	90 mL of 30 mM selenious acid and 1.8 mL of 40 mM ascorbic acid	30 min	RT	–	160–170 rpm	SeNPs	Spherical	30–80 nm	293 nm	Radhika et al. (2022)
Kappaphycus alvarezii							AgNPs	Polymorphic and cuboidal	80 nm		Khan et al. (2022a, b)
Laurencia aldingensis, Laurenciella sp.							AgNPs	Spherical, hexagonal and triangular	5–10 nm	440 nm	Vieira et al. (2016)
Laurencia catarinensis	1 gm of DLE	100 mL of 10^{-3} M $AgNO_3$	3 h	RT		Stirring	AgNPs	Spherical	39.41–77.71 nm	401	Abdel-Raouf et al. (2018)
Laurencia obtusa	30 mg of polysaccharide extract	1 mL of 0.1 M $AgNO_3$	15, 30, 45, and 60 min	90 °C	12		AgNPs	Spherical	5–10 nm	400–420 nm	Zahran and Mohammed (2021)
Laurencia obtusa	10 mL of DLE	90 mL of 1 mM $AgNO_3$	40 min	30 °C	–	Magnetic stirring	AgNPs	–	–	428 nm	Khalifa et al. (2016)
Laurencia papillosa	5 gm of DLE	5 mM $HAuCl_4$		RT			AuNPs	Flat and hopper triangular	3.5–53 nm	526–586 nm	Montasser et al. (2016)

(continued)

1.5 Green Synthesis of MFNPs

Table 1.1 (continued)

Marine plant	*Plant part and concentration	Precursor concentration	Reaction time	Synthesis temperature	Reaction pH	Reaction mixture stirring	Synthesized nanoparticles	Shape	Size range	UV–VIS spectrum	Reference
Laurencia papillosa	10 mL of DLE	90 mL of 1 mM $AgNO_3$	72 h	RT	–	Magnetic stirring	AgNPs	Cubic	–	450 nm	Omar et al. (2017)
Osmundaria obtusiloba	5 mL of DLE	20 mL of 1 mM $HAuCl_4$	2 h	60 °C	–	200 rpm	AuNPs	Spherical, triangular, and diamond	10–20 nm	540 nm	Rojas-Pérez et al. (2015)
Palmaria decipiens	FLE	50 and 150 µL of 0.01 M $HAuCl_4$ and 100–500 µL of 0.005 M $AgNO_3$		RT		Stirring	AuNPs and AgNPs	Drop like and spherical	36.8 ± 5.3 nm and 17.8 ± 2.6 nm	548 nm and 425 nm	González-Ballesteros et al. (2018)
Polysiphonia sp.	6 mL of WE	100 mL of 0.01 mM $AgNO_3$	2 h	RT		Stirring	AgNPs	Spherical	5–25 nm	420 nm	Moshfegh et al. (2019)
Porphyra linearis	0.008 g/mL DLE	0.4 mM $HAuCl_4$	24 h	30 °C	–	–	AuNPs	Spherical	10–22 nm	529 nm	González-Ballesteros et al. (2022)
Porphyra vietnamensis	0.01% (w/v) porphyran	100 mL of 1 × 10^{-4} M $HAuCl_4$	15 min	70 °C	11	–	AuNPs	Spherical	13 ± 5 nm	520 nm	Venkatpurwar et al. (2011)
Porphyra vietnamensis	0.01% (w/v) porphyran	100 mL of 10^{-3} M $AgNO_3$	15 min	70 °C	11	–	AgNPs	Spherical	13 ± 3 nm	404 nm (Omar et al. 2017)	(Venkatpurwar and Pokharkar 2011)
Porphyra vietnamensis	0.01% (w/v) porphyran	100 mL of 1 × 10^{-4} M $HAuCl_4$	15 min	70 °C	11	–	AuNPs	Spherical	14 ± 2 nm	520 nm	Venkatpurwar et al. (2012)
Portieria hornemannii							AgNPs	Spherical	35–50 nm	418 nm	Fatima et al. (2020)
Portieria hornemannii							AgNPs	Spherical	9–80 nm	430 nm	Ramamoorthy et al. (2019)
Portieria hornemannii	20 mL DLE	10 mL of 1 mM $AgNO_3$	30 min	60 °C	–	5000 rpm	AgNPs	Spherical	100–200 nm	430 nm	Sabatini and Anchana Devi (2017)

(continued)

Table 1.1 (continued)

Marine plant	*Plant part and concentration	Precursor concentration	Reaction time	Synthesis temperature	Reaction pH	Reaction mixture stirring	Synthesized nanoparticles	Shape	Size range	UV-VIS spectrum	Reference
Portieria hornemannii (Chondrococcus hornemannii)	5 mL of DLE	95 mL of 10^{-3} M AgNO$_3$	1 h	100 °C	–	–	AgNPs	Spherical	35–75 nm	430 nm	Aravindan et al. (2014)
Pterocladia capillacea	90 mL of crude hot water soluble polysaccharide solution	1 mL of 0.1 mM AgNO$_3$	20 min	70 °C	10	Stirring	AgNPs	Spherical	7 nm	407–424 nm	El-Rafie et al. (2013)
Pterocladia capillacea	0.02 g of DLE	100 mL of 3 mM copper sulfate solution	3 h	70 °C	–	Magnetic stirring	CuONPs	Irregular	62 ± 17.7 nm	290–330 nm	Aboeita et al. (2022)
Pterocladiella capillacea							AgNPs	Spherical	4–80 nm	430 nm	Negm et al. (2018)
Pterocladiella capillacea	10 m of DLE	90 mL of 1 mM AgNO$_3$	48 h	RT	–	120 rpm	AgNPs	Spherical	11.4 ± 3.52 nm	400 nm	Kassas and Attia (2014)
Pterocladiella capillacea	DLE	0.1 M FeCl$_3$	2 h	RT		800 rpm	Fe$_3$O$_4$NPs	Nanospheres	16.85–22.47 nm	415 nm	Salem et al. (2019)
Pyropia yezoensis							AgNPs	Spherical	20–22 nm	450 nm	Ulagesan et al. (2021)
Rhodymenia palmata	DLE	1 mM AgNO$_3$	10 min	–	–	–	AgNPs	Spherical	50 nm	338–490 nm	Murugammal and Flora (2017)
Solieria robusta	20 mL DLE	0.01 M AgNO$_3$	24 h	RT	–	–	AgNPs	Cubic and cylindrical	17 nm	420 nm	Khan et al. (2022a, b)
Spyridia filamentosa	20 mL DLE	80 mL of 1 mM AgNO$_3$	2 h	36 °C	–	Static	AgNPs	Spherical	20–30 nm	420 nm	Valarmathi et al. (2020)
Spyridia fusiformis	10 mL of DLE	90 mL of 1 mM AgNO$_3$		RT		Stirring	AgNPs	Spherical	5–50 nm	450 nm	Murugesan et al. (2017)
Spyridia hypnoides							AgNPs	Spherical	49 nm	412 nm	Roseline et al. (2019)
Seagrasses											
Amphibolis antarctica											Fawcett et al. (2017)

(continued)

1.5 Green Synthesis of MFNPs

Table 1.1 (continued)

Marine plant	*Plant part and concentration	Precursor concentration	Reaction time	Synthesis temperature	Reaction pH	Reaction mixture stirring	Synthesized nanoparticles	Shape	Size range	UV-VIS spectrum	Reference
Cymodocea serrulata							AgNPs	Spherical	17–29 nm	430 nm	Chanthini et al. (2015)
Cymodocea serrulata	5 mL of DLE	95 mL of 1 mM AgNO$_3$	1 h	4 °C, 60 °C, and RT	-	-	AgNPs	Spherical	5–25 nm	420 nm	Palaniappan et al. (2015)
Cymodocea serrulata	1 gm of DLE	100 mL of 10^{-3} M AgNO$_3$	16 h	37 °C		60 rpm	AgNPs	Spherical	40.49–66.44 nm	431 nm	(RathnaKumari et al. 2018)
Cymodocea serrulata	12 mL of DLE	88 mL of 1 mM AgNO$_3$	–	RT	–	Stirring	AgNPs	Spherical	56–87 nm	415 nm	Amutha et al. (2019)
Cymodocea serrulata	12 mL of DLE	88 mL of 1 mM PdCl$_2$	–	RT	–	Stirring	PdNPs	Spherical	91–98 nm	370 nm	Amutha et al. (2019)
Cymodocea serrulata	12 mL of DLE	88 mL of 1 mM TiO(OH)$_2$	–	RT	–	Stirring	TiO$_2$ NPs	Spherical and aggregated	62–87 nm	360 nm	Amutha et al. (2019)
Cymodocea serrulata	5 mL of DLE	95 mL of 1 mM AgNO$_3$	1 h	60 °C	–		AgNPs	Spherical	5–80 nm	423 nm	(Kailasam et al. 2023)
Enhalus acoroides							AgNPs	Poly-dispersed and spherical	2–100 nm	419 nm	(Senthilkumar et al. 2016)
Halophila stipulacea	10 mL of DLE	90 mL of 1 mM AgNO$_3$	48 h	RT	–	120 rpm	AgNPs	Spherical	17.7–25 nm	~402 nm	El-Kassas and Ghobrial (2017)
Halophila stipulacea	DLE	1:1 ratio of 0.1 M FeCl$_3$	60 min	RT	–	120 rpm	Fe$_3$O$_4$ NPs	Spinal and needle	17.7–25 nm	~402 nm	El-Kassas and Ghobrial (2017)
Posidonia australis					4		AuNPs				Fawcett et al. (2017)
Syringodium isoetifolium	5 mL of DLE	95 mL of 1 mM AgNO$_3$	48 h	45 °C	–		AgNPs	Polydispersed and spherical	2–50 nm	422 nm	Ahila et al. (2016)
Xylocarpus granatum							AgNPs	–	–	426 nm	Maity and Mondal (2017)
Mangroves											

(continued)

Table 1.1 (continued)

Marine plant	*Plant part and concentration	Precursor concentration	Reaction time	Synthesis temperature	Reaction pH	Reaction mixture stirring	Synthesized nanoparticles	Shape	Size range	UV-VIS spectrum	Reference
Acanthus ilicifolius	10 mL of FLE	90 mL of AgNO$_3$ (21.2 gm of AgNO$_3$ powder in 125 mL of Milli Q water)	10 min	RT	–	–	AgNPs	Spherical	180 nm	420 nm	Ali et al. (2015)
Avicennia alba							AgNPs	Spherical and cuboidal	25.3–65.4 nm	448 nm	Bakshi et al. (2015)
Avicennia alba	20 mL of DLE	80 mL of 1 mM AgNO$_3$	15 min	60 °C	–	–	AgNPs	Different shapes	–	455 nm	Nagababu and Rao (2016)
Avicennia marina	10 mL of FLE	90 mL of AgNO$_3$	8 h	RT			AgNPs	Spherical	71–110 nm	420 nm	Gnanadesigan et al. (2012)
Avicennia marina	10 mL of FLE	90 mL of AgNO$_3$	10 min	RT	–	–	AgNPs	–	60–95 nm	420 nm	Balakrishnan et al. (2016)
Avicennia marina	DLE	1 mL of AgNO$_3$		RT			AgNPs	Spherical to rectangular	10–40 nm	316 nm	Barnawi et al. (2019)
Avicennia marina							AgNPs	Spherical	10–75 nm	420 nm	Abdi et al. (2018)
Avicennia marina	10 mL of seed extract	90 mL of 6 mM AgNO$_3$	8 h min	RT			AgNPs	Triangular, hexagonal, and spherical	5–10 nm	420 nm	Naidu et al. (2019)
Avicennia marina	1 gm of DLE	Titanium isopropoxide	150 min	37 °C	–	–	TiO$_2$ NPs	Spherical	25–250 nm	400 nm	Lefteh et al. (2020)
Avicennia marina	10 mL of FLE	90 mL of 1 mM AgNO$_3$	2 h	37 °C	–	–	AgNPs	Spherical	10–20 nm	400 nm	Tian et al. (2020)
Avicennia officinalis	DLE	10 mM AgNO$_3$	6 h	RT	–	–	AgNPs	Poly-dispersed	50–1000 nm	420 nm	Das et al. (2019)
Ceriops decandra	10 mL of FLE	90 mL of AgNO$_3$	10 min	RT	–	–	AgNPs	Spherical	–	400 nm	Sankar and Abideen (2019)
Ceriops decandra	20 mL of 0.5 mg/mL of polysaccharide	20 mL of 1 mM AgNO$_3$	90 min	RT	–	Stirring	AgNPs	Spherical	28 nm	426 nm	Maity et al. (2019)
Ceriops tagal	5 mL of SE	95 mL of 1 mM aqueous CuSO$_4$		RT			CuNPs		~4 nm	500 nm	Ramteke et al. (2018)

(continued)

1.5 Green Synthesis of MFNPs

Table 1.1 (continued)

Marine plant	*Plant part and concentration	Precursor concentration	Reaction time	Synthesis temperature	Reaction pH	Reaction mixture stirring	Synthesized nanoparticles	Shape	Size range	UV–VIS spectrum	Reference
Ceriops tagal	DLE	0.008 M AgNO₃	–	RT	–	130 rpm	AgNPs	Spherical	30 nm	424 nm	(Dhas et al. 2013)
Excoecaria agallocha	2 mL of DLE	20 mL of 1 mM AgNO₃	5 h	–	–	–	AgNPs	Spherical and hexagonal	23–42 nm	440 nm	Bhuvaneswari et al. (2017)
Excoecaria agallocha	FLE						Ag-ZnONPs	Spherical and triangular	~90–100 nm	380 nm	(Khan et al. 2020)
Excoecaria agallocha	FLE						ZnONPs	Spherical	90–100 nm	200 nm	Khan et al. (2020)
Excoecaria agallocha	10 mL of FLE	190 mL of 1 × 10⁻³ M AgNO₃	30 min	RT			AgNPs	Spherical	18–50 nm	440 nm	Kumar et al. (2016)
Excoecaria agallocha	1–5 mL of DLE	100 mL of 1 mM AgNO₃	35 min		–	–	AgNPs	Spherical	15–43 nm	434 nm	Arun et al. (2014)
Excoecaria agallocha											Spadling (1997)
Heritiera fomes	10 mL of DLE	50 mL of 1 mM AgNO₃	12 h	RT	–	Stirring	AgNPs	–	50 nm	375 nm	Thatoi et al. (2016)
Heritiera fomes	10 mL of DLE	50 mL of 0.5 M ZnCl₂	12 h	RT	–	Stirring	ZnONPs	–	40–50 nm	350 nm	Thatoi et al. (2016)
Lumnitzera racemosa	FLE	3 mL of a 0.01 M AgNO₃		RT			AgNPs			537 nm	Arshan et al. (2020)
Rhizophora apiculata							AgNPs	Spherical	19–42 nm	422 nm	Antony et al. (2011)
Rhizophora apiculata							AgNPs	Irregular	35–100 nm	458.9 nm	Alsareii et al. (2022)
Rhizophora lamarckii	3 mL of DLE	47 mL of 1 mM AgNO₃	6 h	–	–	–	AgNPs	Polydispersed spherical	12–28 nm	420 nm	Kumar et al. (2017a, b)
Rhizophora mucronata	10 mL of FLE	100 mL of 1 mM AgNO₃	5 min	121 °C & 15 psi	–	–	AgNPs	Spherical	4–26 nm	426 nm	Umashankari et al. (2012)
Rhizophora mucronata	DLE	0.001 M AgNO₃	24 h	–	5.5	120 rpm	AgNPs	Spherical	5.8–9.21 nm	460 nm	Manoj Singh et al. (2013a, b)
Rhizophora mucronata							AgNPs	Spherical	1–80 nm	420 nm	Abdi et al. (2019a)

(continued)

Table 1.1 (continued)

Marine plant	*Plant part and concentration	Precursor concentration	Reaction time	Synthesis temperature	Reaction pH	Reaction mixture stirring	Synthesized nanoparticles	Shape	Size range	UV–VIS spectrum	Reference
Rhizophora mucronata	10 mL of FLE	90 mL of DLE	10 min	RT	–	–	AgNPs	Spherical	60–95 nm	420 nm	Gnanadesigan et al. (2011)
Rhizophora mucronata	50 mL of DLE	50 mL of 10^{-3} mM AgNO$_3$	15 h			Stirring	AgNPs	Cubic	25 nm	421 nm	Premanathan et al. (2014)
Rhizophora mucronata	10 mL of FLE	90 mL of AgNO$_3$	10 min	RT	–	–	AgNPs	Spherical	–	430 nm	Sankar and Abideen (2019)
Rhizophora mucronata	10 mL of BE, DLE, and SE	90 mL of 20 mM AgNO$_3$	90 min	RT	–	Stirring	AgNPs	Spherical	1–80 nm	420 nm	Abdi et al. (2019b)
Rhizophora stylosa							AgNPs	Spherical	5–87 nm	403–443 nm	Willian et al. (2022)
Rhizophora stylosa	DLE	10 mM AgNO$_3$	3 months	RT			AgNPs	Spherical	9–57 nm	439–453 nm	Willian et al. (2020)
Sonneratia alba	DLE	50 mL of 1 mM AgNO$_3$	72 h	25 °C	–	–	AgNPs	Spherical and cubical	20–60 nm	421 nm	Murugan et al. (2017)
Sonneratia apetala							AgNPs	Spherical and cuboidal	33.1–48.1 nm	419 nm	Bakshi et al. (2015)
Sonneratia apetala	20 mL of DLE	80 mL of 1 mM AgNO$_3$	–	60 °C	–	–	AgNPs	Spherical	–	425 to 475 nm	Nagababu and Rao (2017)
Sonneratia apetala	10 mL of DLE	50 mL of 1 mM AgNO$_3$	12 h	RT	–	Stirring	AgNPs	–	20–30 nm	383 nm	Thatoi et al. (2016)
Sonneratia apetala	10 mL of DLE	50 mL of 0.5 M ZnCl$_2$	12 h	RT	–	Stirring	ZnONPs	–	70–100 nm	354 nm	Thatoi et al. (2016)
Sonneratia caseolaris							AgNPs	Spherical and cuboidal	18.3–53.2 nm	424 nm	Bakshi et al. (2015)
Xylocarpus granatum	BE	10 mM AgNO$_3$	6 h	RT	–	–	AgNPs	Poly-dispersed	20–1000 nm	470 nm	Das et al. (2019)
Xylocarpus granatum							AuNPs	Spherical	17 nm	–	Pisitsak et al. (2021)

(continued)

1.5 Green Synthesis of MFNPs

Table 1.1 (continued)

Marine plant	*Plant part and concentration	Precursor concentration	Reaction time	Synthesis temperature	Reaction pH	Reaction mixture stirring	Synthesized nanoparticles	Shape	Size range	UV–VIS spectrum	Reference
Sea-blites and other plants associated with marine plants											
Citrullus colosynthis	10 mL of fresh callus exctract	90 mL of 1 mM AgNO₃	24 h	RT	–	–	AgNPs	Spherical	85–100 nm	–	Satyavani et al. (2011)
Hibiscus tiliaceus	10 mL of FLE	190 mL of 1 mM AgNO₃	3 h	RT	–	–	AgNPs	Spherical	20–65 nm	420 nm	Usha Rani et al. (2016)
Ipomoea pes-caprae	20 mL of DRE	1 mM AgNO₃	3 h	50 °C	–	Stirring	AgNPs	Spherical	50 nm	400 nm	Subha et al. (2015)
Ipomoea pes-caprae	10 mL of fresh callus exctract	90 mL of 1 mM AgNO₃	24 h	RT	–	–	AgNPs	Rectangular and irregular	2–45 nm	330 nm	Satyavani et al. (2013)
Ipomoea pes-caprae	20 mL of DRE	1 mM AgNO₃	3 h	50 °C	–	Stirring	AgNPs	Spherical	< 100 nm	448 nm	Veeramani et al. (2018)
Suaeda monoica	10 mL of DLeE	90 mL of 1 mM AgNO₃	5 h	35 °C	–	–	AgNPs	Spherical	31 nm	430 nm	Satyavani et al. (2012)
Sesuvium portulacastrum	5 mL of FLE	45 mL of 10^{-3} M AgNO₃	24 h	RT	–	–	AgNPs	Spherical	5–20 nm	430 nm	Kathiresan et al. (2012)
Sesuvium portulacastrum	5 mL of FLE	45 mL of 10^{-3} M AgNO₃	24 h	RT	–	–	AgNPs	Spherical	5–20 nm	420 nm	Nabikhan et al. (2010)
Clerodendrum inerme	5 mL of FLE	45 mL of 10^{-3} M AgNO₃	24 h	RT	–	–	AgNPs	Spherical	5–20 nm	430 nm	Kathiresan et al. (2012)

BE: Bark extract; DLE: Dried leaf/thallus/frond extract; FLE: Fresh leaf/thallus extract; HM: Hand mixing; DRE: Root/rhizome extract; RT: Room temperature; SE: Stem extract; and WE: Whole extract. *Plant part used to make aqueous extract for nanoparticle synthesis

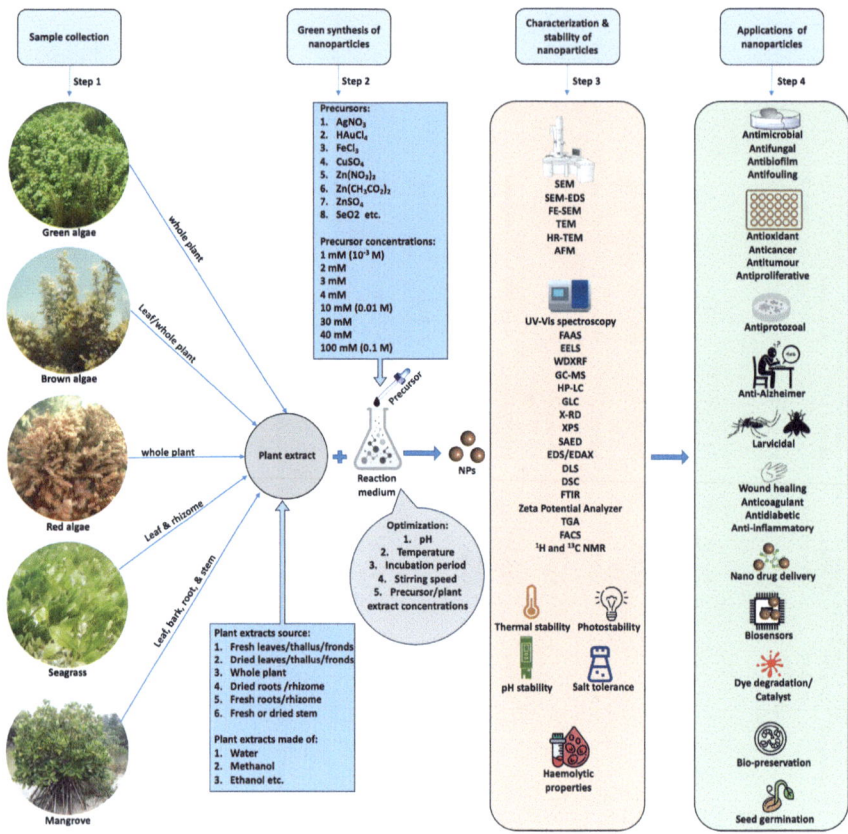

Fig. 1.2 Illustration depicting methodology flow involved in green synthesis of nanoparticles from marine plants

The collected marine plants are first washed with fresh water (tap water) thoroughly to remove the extraneous substances such as detritus, sand, sand pebbles, shells, epiphytes, settled organisms, other impurities, salts, and other chemical constituents on the surface. A study also suggested use of betadine solution and antibiotic mixtures to remove surface contaminants from algal biomass overnight (Parial et al. 2012). In case of gold nanoparticle synthesis, impurities adsorbed on samples were washed with dilute HCl and deionized water to avoid their interference during the formation of nanoparticles (Oza et al. 2012). Then, the plants are cleaned (2–3 times) again in distilled water using a brush to remove hardly bound particles and debris, and then placed on blotting paper to remove remaining water drops. These cleaned marine plants are dried (shade dry for 3–7 days; oven dry for overnight or 24–48 h at 37 °C – 70 °C; lyophilized; freeze dried in liquid nitrogen at – 196 °C for few seconds to minutes or at – 20 °C or – 80 °C for several hours), pulverized/

1.5 Green Synthesis of MFNPs

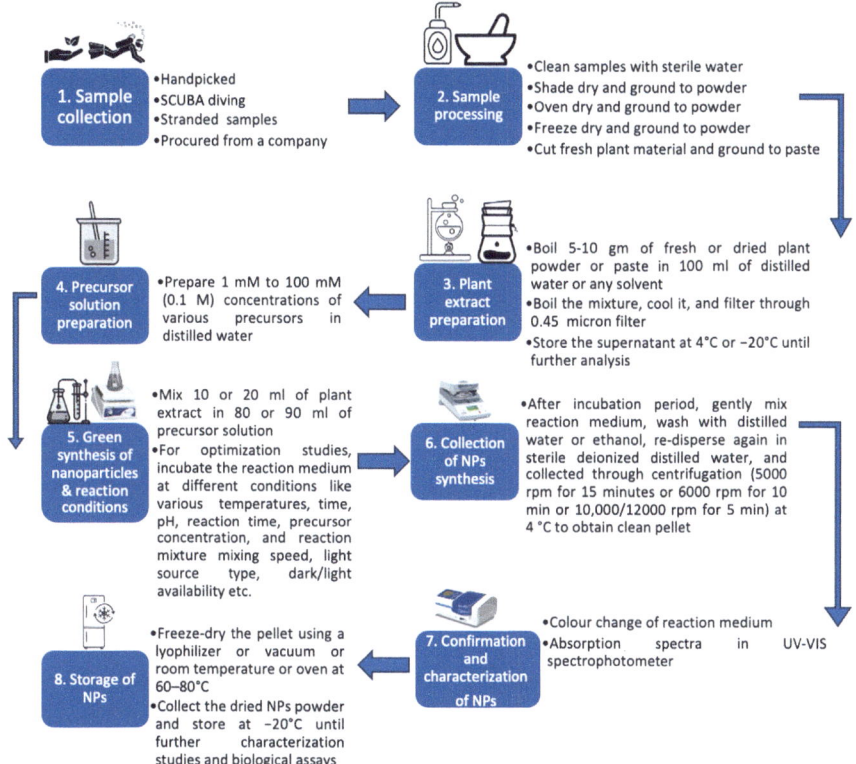

Fig. 1.3 Illustration depicting a detailed flow chart of methodology involved in green synthesis of nanoparticles from marine plants. Icons are obtained from Iconfinder.com for illustration purpose

grounded to fine powder using motor pestle or mixer/electric blender, sieved using < 0.5 mm mesh, and stored in refrigeration at 4 °C or − 20 °C until further analysis.

The stored powder from sample is collected (1 gm), mixed with interested solvent (100 mL) (water, methanol, ethanol, etc.) for 10 min–40 h at room temperature or at desired temperature, and filtered through muslin cloth or nylon membrane filters 0.2 μm or Whatman paper No. 1 and centrifuged (8000 rpm for 10 min) to remove residual debris and obtain a clear solution of aqueous extract. Alternatively, the washed plants are chopped into fine pieces and boiled in 100 mL of distilled water in an Erlenmeyer flask, cooled and centrifuged (5000 rpm for 5 min at 4 °C or 10,000 rpm for 30 min at 4 °C) or filtered through muslin cloth or Whatman paper No. 1 or membrane filter 0.43 micron, to obtain a clear solution of aqueous extract. Or, fresh leaves are were macerated in interested solvent (e.g., distilled water, Tris–Cl, solvent etc.) using an agate mortar and pestle, centrifuged, collected supernatant, transferred to fresh sterile centrifuge tubes, and stored at 4 °C. The obtained aqueous extract is either used instantly to synthesis nanoparticles or stored at − 20 °C for longer period until synthesizing nanoparticles. Fresh plants can also be cut into small pieces, grounded to fine paste using mortar and pestle, dissolved in 100 mL of distilled water, boiled at 60 °C for 10–15 min (or sonicated for 30 min to obtain aqueous extract), allowed cool at room temperature, filtered through Whatman filter paper and used the aqueous extract for synthesizing nanoparticles. In most case nanoparticles synthesized extracellularly, but in some cases, algal nanoparticles are synthesized intracellularly (Parial et al. 2012; Dahoumane et al. 2017; Khanna et al. 2019), which are extracted by sonicating and purification as detailed above.

The aqueous extracts (20 mL) are supplemented with drops of interested precursor solution (80 mL) at specified parameters (e.g., various temperatures, pH (pH of precursor solution is adjusted using 0.1 M HCl and 0.1 M NaOH), reaction time, precursor concentration, and reaction mixture mixing speed) for synthesizing nanoparticles. These aqueous extracts in fact serve as reducing and stabilizing agents to precursor solution as well as capping agents to synthesized nanoparticles. The synthesized nanoparticles in the reaction mixture are gently mixed, carefully washed with distilled water to remove biological matter and debris, re-dispersed again in sterile deionized distilled water, and collected through centrifugation (5000 rpm for 15 min or 6000 rpm for 10 min or 10,000/12000 rpm for 5 min) at 4 °C to obtain clean pellet. The pellet is freeze-dried using a lyophilizer or vacuum dried or oven dried at 60–80 °C, then collected and stored at − 20 °C until further analysis or used immediately for characterization studies and biological assays. Alternatively, the pellet is washed with ethanol and oven dried at 80 °C for 2 h (Vikneshan et al. 2020) or air dried at room temperature for 3–4 weeks to obtain nanoparticles in powder form (Khan et al. 2022a, b). In case of MgONPs (Pugazhendhi et al. 2019) and ZnONPs (Alsaggaf et al. 2021), pure nanoparticles are obtained by calcinating the mixture at 400–500 °C and 455 °C for 3–4 h and 4 h, respectively, in a muffle furnace. In case of AuNPs, dialysis was performed for 24 h using 12 kDa dialysis tubing to remove ionic impurities and unreacted reducing oligosaccharides from the final product (Venkatpurwar and Pokharkar 2011; Venkatpurwar et al. 2011; Chen et al. 2018). The epirubicin-loaded kappa-carrageenan capped gold nanoparticles dispersion was dialyzed using 14 kDa dialysis membrane to remove free EPI and CAO-AuNPs (Chen et al. 2019). Similarly, doxorubicin loaded acetylated fucoidan nanoparticles were obtained upon dialysis against deionized water (Lee et al. 2013). The pellet or powdered form is then stored in amber colored air-tight containers or packets until further analysis (Fig. 1.3).

1.6 Morphology and Identification of Nanoparticles

The formation of green synthesized nanoparticles is usually observed within 10 min to 48 h, resulting a wide range of morphological forms of nanoparticles (Fig. 1.4; Table 1.1). A detailed method of synthesizing nanoparticles from marine flora is detailed in supplementary information (Fig. 1.3). Most of the metal and metal-oxide nanoparticles derived from marine plants possess spherical shape, followed by hexagonal, triangular, cubic/cuboidal, and oval/ellipsoid structures (Table 1.1). Based on

1.6 Morphology and Identification of Nanoparticles

the literature, nanoparticles originated from seaweed extracts were found to form diverse structures than nanoparticles originated from seagrass, mangrove, and other coastal plant extracts. exceptionally, AgNPs derived from extracts of green seaweed *Chaetomorpha linum* (R. Ragupathi Raja Kannan et al. 2013a, b), brown seaweed, *Padina pavonica* (Abdel-Raouf et al. 2019), red seaweeds, *Chondrus crispus* (Castro et al. 2013), *Kappaphycus alvarezii* (Khan et al. 2022a, b), *Solieria robusta* (K. D. Khan et al. 2022a, b), seagrass *Halophila stipulacea*, (El-Kassas and Ghobrial 2017), mangroves, *Avicennia alba* (Nagababu and Rao, 2016), *Avicennia marina* (Barnawi et al. 2019), *Rhizophora apiculata* (Alsareii et al. 2022), displayed nano-clusters coalescence, polygonal, wire like, polymorphic, cylindrical, spinal/needle, varied, rectangular, and irregular shaped nanoparticles, correspondingly. Polygonal shape AgCl nanoparticles were observed from *Ulva fasciata* exctrates (Lashgarian et al. 2021).

Gold nanoparticles derived from brown seaweeds extracts showed shapes such as truncated triangular from *Sargassum* sp. (Liu et al. 2005), octahedral from *Sargassum bovinum* (Momeni and Nabipour 2015), truncated octahedral from *Padina gymnospora* (V. Sri Ramkumar et al. 2017a, b), pentagons from *Padina gymnospora* (Singh et al. 2015). AuNPs fabricated from the extracts of some red seaweeds demonstrated a range of morphological nanostructures such as polygonal and truncated triangular from *Actinotrichia fragilis* (Momeni et al. 2016), quasi-spherical from *Chondrus crispus* (Alvarez-Vinas et al. 2022), truncated triangular and hexagonal from *Galaxaura elongata* (Abdel-Raouf et al. 2017), hopper triangular structures from *Laurencia papillosa* (Montasser et al. 2016), flower structures on triangular, truncated triangular, and truncated pentagon NPs from *L. papillosa* (Montasser et al. 2016), polyhedral from *Gelidium corneum* (González-Ballesteros et al. 2022), rectangular from *Halymenia pseudofloresii* (Palaniyandi et al. 2023), pentagons from *Hypnea musciformis* (Vadlapudi and Amanchy 2017), and diamond shape from *Osmundaria obtusiloba* (Rojas-Pérez et al. 2015).

Flower-like nanoparticles were reported only for metal-oxide nanoparticles derived from green seaweed, *Ulva fasciata* (ZnONPs) (Alsaggaf et al. 2021), brown seaweed, *Sargassum wightii* (MgONPs) (Pugazhendhi et al. 2019), and red seaweed, *Gracilaria gracilis* (ZnONPs) (Francavilla et al. 2014). Small fibrillar pattern nanoparticles were observed from CdO-ZnONPs synthesized from brown seaweed *Padina gymnospora* (Rajaboopathi and Thambidurai 2017). Square shaped ZnNPs were reported from extracts of brown seaweed *Sargassum latifolium* (El-Khateeb et al. 2019).

The color change in the solution during the nanoparticle's synthesis process indicates the formation of nanoparticles (Table 1.1). AgNPs are often found in light yellow to dark brown color. Au-NPs can form pale yellow, pinkish, ruby red, red, or pinkish colors. Zinc Oxide nanoparticles (ZnONPs) are identified by color change from green to pale white (Fouda et al. 2022b). Palladium nanoparticles are identified by color change from yellow to dark brown. Similarly, platinum nanoparticles display light yellow to dark yellowish-brown color. CdSNPs (Sujitha et al. 2017), CdO-ZnONPs (Rajaboopathi and Thambidurai 2017), DOX-AcFuNPs (Lee et al. 2013), FCD/LFNPs (Etman et al. 2020), FCD/QCNPs (Etman et al. 2021),

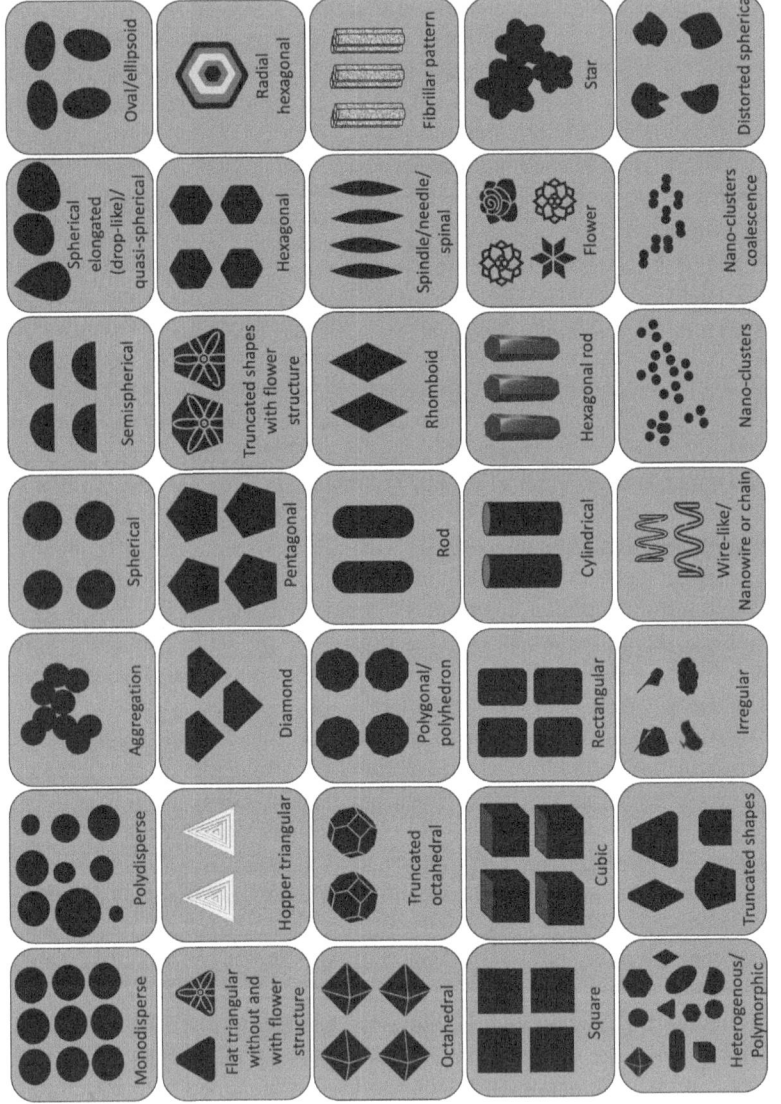

Fig. 1.4 Morphology of various nanoparticles derived from marine plant extracts

Fu/CHNPs (Oliveira et al. 2018), ULANP (Al-Malki 2020), PEI-FCD-DOX NPs (Pawar et al. 2019), and Zinc nanoparticles (ZnNPs) (El-Khateeb et al. 2019) color details are unavailable in the literature. Copper nanoparticle (CuNPs) turn to brick red from blue solution. Whereas, copper oxide nanoparticles (CuNPs) are confirmed by reaction mixture color change from green to brown (Rajeshkumar et al. 2021). The confirmation of cobalt oxide nanoparticles (CoONPs) formation is indicated by a transition of initial pale pink colored reaction mixture to a dark brownish hue (Ajarem et al. 2022). Palladium nanoparticles (PdNPs) synthesis is confirmed by the color change from light yellow to dark brown (Sonbol et al. 2021). TiO_2NPs synthesis is noticed by color transition of dark brown reaction mixture to light brown color (Lefteh et al. 2020). MgONPs demonstrate color change from pale yellow to yellowish-brown (Fouda et al. 2022a). FeONPs displays light brown, brown, and reddish-brown color (Salem et al. 2020). Doxorubicin (DOX) loaded Protamine/fucoidan NPs show a milky white color (Lu et al. 2017). A successful synthesis of epirubicin-loaded kappa-carrageenan gold nanoparticles (EPI-CAO-AuNPs) is indicated by orange red color (Chen et al. 2019). SeNPs forms ruby red to brick red color (Radhika et al. 2022). Platinum nanoparticles (PtNPs) indicated by a color change from light yellow to a dark yellowish-brown hue (Shiny et al. 2014). Superparamagnetic iron oxide nanoparticles (SPIONPs) show a color change from dark brown to dark black (Khaleelullah et al. 2017). AgClNPs formation is identified by yellowish-brown color (Dhas et al. 2014a, b). Zirconium oxide nanoparticles (ZrONPs) synthesis is indicated by a white flame during combustion at 400 °C (Kumaresan et al. 2018).

The identification of nanoparticles is relatively easy compared to other marine natural products. The identification of synthesized nanoparticles is in general confirmed by: (1) the color change of solution, (2) absorption spectra confirmation with UV–VIS spectroscopy, (3) and distribution, hydrodynamic diameter, and surface charge of nanoparticles with dynamic light scattering (DLS) spectroscopy, (4) stability of nanoparticles, nanoconjugates (e.g., Ag-Thymol), and bimetallic nanoparticles (e.g., Ag-Au), using zeta potential, (5) morphological (shape and size) analysis with SEM or TEM image analysis, (6) elemental presence by Energy dispersive X-ray analysis (EDX), (7) crystallinity, crystal structure, and lattice parameters with selected area electron diffraction (SAED) pattern, (8) identification of functional groups using FTIR, and (9) purity, crystalline nature, geometry, crystalline size, phases, orientation, and chemical composition with XRD analysis.

1.7 Factors Affecting the Synthesis of Nanoparticles

The successful synthesis of nanoparticles from marine plants depends on various factors. The shape and size of nanoparticles synthesized from microbes and plant sources was observed to be affected by various factors (Fig. 1.5) like reaction time, precursor concentration, pH, temperature, light, type of biomolecule (phytochemical components) present in source of synthesis, and biomass/aqueous extract concentration (Chugh et al. 2021). Specially, pH and temperature are two important factors known to impact the synthesis and production of nanoparticles (Hashemi et al. 2015). Considerably, the formation of AgNPs fabricated from *Caulerpa serrulata* was observed best at high temperature at 95 °C and higher alkaline pH of 9.95 than acidic pH 4.10 (Aboelfetoh et al. 2017). Gold nanoparticles size and shape were affected by temperature and pH, where high temperature (75 °C) found to produce smaller size nanoparticles compared to lower (30 °C) temperature, and varied shapes were detected at 25 and 60 °C (Singh et al. 2015). High amount of AgNPs formation from silver ions was achieved at 100 °C using the aqueous seaweed extracts of *Sargassum cinereum* as reducing agent (Mohandass et al. 2013). Increasing calcination temperature from 400 to 600 °C was reported to reduce ZnONPs size from 31.4 to 14.7 nm (Sari et al. 2017). Likewise, increasing plant extract concentration from 5 to 20% resulted in increased particle size of ZnONPs from 607 to 649 nm (Sari et al. 2017). The synthesis process of AgNPs and Zinc oxide nanoparticles from seaweeds were triggered by pH, temperature, concentrations of seaweed extracts and precursor (Ghaemi and Gholamipour 2017; Nagarajan and Kuppusamy, 2013). Increased temperatures reportedly caused reduction of AgNPs size (Anuluxan et al. 2022). Significant decrease in size was observed at 90 °C for Zinc oxide nanoparticles (Nagarajan and Kuppusamy 2013). Temperature known to play important role in agglomeration and size of nanoparticles (Prasad et al. 2013; Vinoth et al. 2019). Calcination of ZrO_2 nanoparticles was reported to enhance antimicrobial activity (Kumaresan et al. 2018).

Nanoparticles with smaller size and larger surface area appear to display high bactericidal properties (Baker et al. 2005). Altering the concentration of the sample extracts as well as metal salts provide synergetic effect of the parameters on the reduction, stabilization, and formation of nanoparticle (González-Ballesteros et al. 2021). The source of algae, size of nanoparticles (e.g., AgNPs), and type of test material (e.g., different fabrics treated with AgNPs) together play a role in deciding the biological properties of nanoparticles like antimicrobial activities (El-Rafie et al. 2013). Thus, optimization studies are imperative in nanoparticles synthesis research (Chen et al. 2018; Costa et al. 2020), to understand the individual and synergetic effect of various physicochemical parameters in synthesizing the efficient nanoparticles from known and new biological sources as well as to investigate effective biological activities.

1.8 Mechanism of Nanoparticle Synthesis from Marine Flora

The mechanisms involved in intracellular and extracellular nanoparticles have been reported only for some nanoparticles. Herein, a brief note is detailed on mechanisms of synthesis for known nanoparticles. Various biomolecules present in the

1.8 Mechanism of Nanoparticle Synthesis from Marine Flora

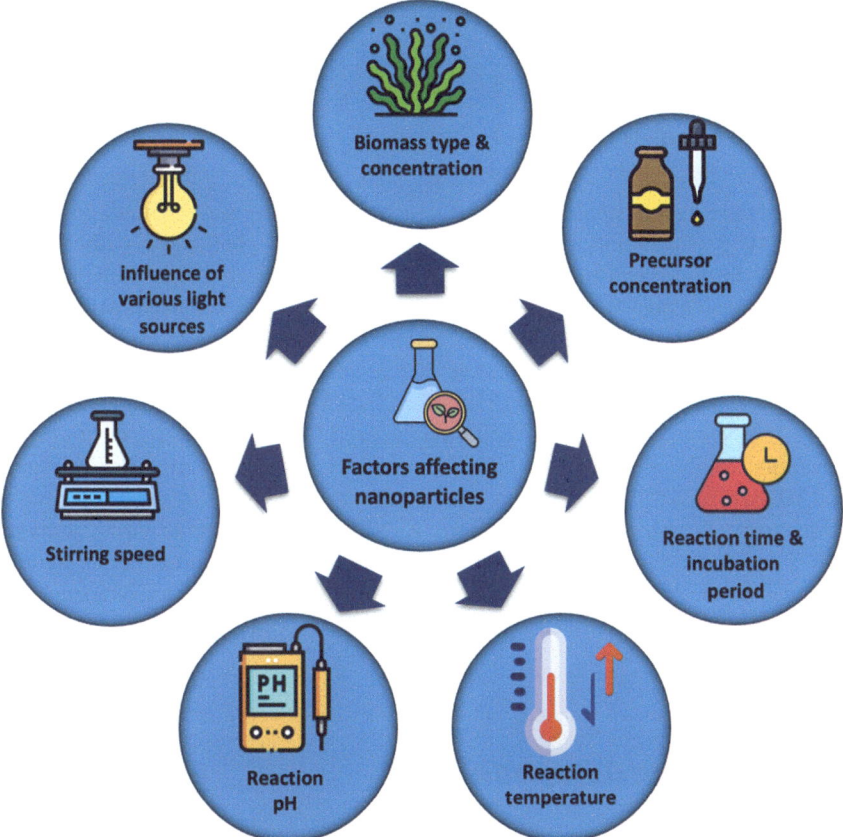

Fig. 1.5 Various factors that affect synthesis of nanoparticles from marine plant extracts. Icon credits: https://www.flaticon.com

plant extracts provide electrons to Ag^+ and transformed to Ag^0 to form AgNPs. The reaction medium become acidic during this process due to released hydrogen ions (Choudhary et al. 2022). Chaudhary et al. (2020) has reviewed the mechanisms involved in synthesizing extracellular and intracellular AuNPs (Chaudhary et al. 2020). While, two possible mechanisms have been proposed for ZnONPs (Alsaggaf et al. 2021). In the first mechanism, algal biomolecules and zinc ions (Zn2+) are chelated to create complexes that are calcined to produce ZnONPs. Whereas, the second mechanism suggest that algal chemical compounds convert zinc ions to zinc metal (Alsaggaf et al. 2021). The synthesis mechanisms of other nanoparticles requires further investigation.

1.9 Applications of Marine Flora-Derived Nanoparticles (MFNPs)

1.9.1 Antiviral Activity

AgNPs obtained from green seaweed *Caulerpa sertularioides* when administered (1.5 mg/10 g concentration) to *Litopenaues vannamei* challenged with WSSV showed highest survival rate of 69.3 ± 1.70% (Anjali et al. 2022). SeNPs fabricated from brown alga *Polycladia myrica* showed antiviral activity against HAV-10, Adenovirus, and HSV-2, with 40.25 ± 2.61%, 8.64 ± 0.82%, and 17.39 ± 1.45% inhibition, respectively (Touliabah et al. 2022). SeNPs synthesized from *P. myrica* also showed moderate inhibition (35.25% ± 0.61%) of HSV-1 (EC_{50} = 52.81 ± 1.03) (Abo-Neima et al. 2023). AgNPs synthesized from red seaweed, *Centroceras clavulatum*, inhibited (>80%) the replication of dengue virus in Vero cells (Murugan et al. 2016a).

Mangrove *Sonneratia alba*-fabricated AgNPs inhibited dengue virus serotype DEN-2 by down-regulating envelope gene and protein at a dose of 15 µg/mL (Murugan et al. 2017). AgNPs synthesized from dried leaf powder of mangrove *Rhizophora lamarckii* Inhibited HIV-1 RTase with IC_{50} of 0.4 µg/mL (S. D. Kumar et al. 2017a, b).

1.9.2 Antibacterial Properties

Nanoparticles have proven to hold significant potential as an innovative resource for alternative treatments distinct from traditional drugs and medications. The mechanism of bacterial inhibition by nanoparticles like AgNPs is triggered by loss of membrane permeability, cell wall rupture, DNA damage, protein synthesis inhibition, increased ROS levels, inhibition of enzymes, and cell cycle arrest (Choudhary et al. 2022). Cobalt oxide nanoparticles also reported to cause DNA damage, altering cell wall permeability, cell wall rupture, and enhancement of ROS levels, which leads to cell death (Ajarem et al. 2022). The similar process is applicable for other nanoparticles that display a wide range of bioactive nature against various microbes (Fig. 1.6; Table 1.2). A wide range of nanoparticles derived from various marine plants reported in the literature have inhibited numerous human pathogens. This book presents only the nanoparticles exhibiting significant inhibitory activity. For comprehensive details on biological properties of other nanoparticles, readers are directed to the supplementary information (Table 1.2).

AgNPs are widely explored from marine floral components than other nanoparticles (Fig. 1.7). These nanoparticles derived from green seaweeds such as *Caulerpa sertularioides* inhibited *Vibrio harveyi*, *V. vulnificus*, and *V. parahaemolyticus*, at 100 µg/mL concentration and displayed 22.96 ± 0.6 mm, 20.14 ± 0.6 mm, and

1.9 Applications of Marine Flora-Derived Nanoparticles (MFNPs)

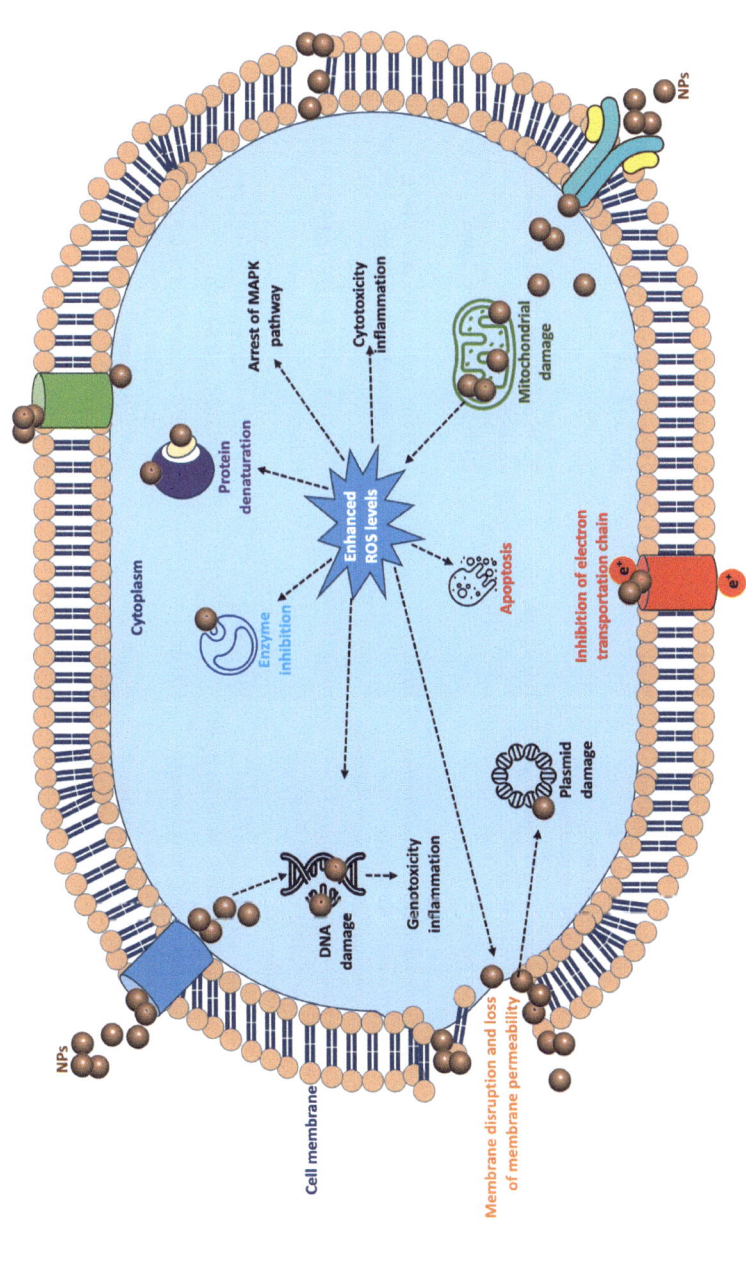

Fig. 1.6 Mechanism of antimicrobial and anticancer activity demonstrated by NPs fabricated from marine plants. Illustration is created based on the following literature (Ajarem et al. 2022; Baskar et al. 2023; Choudhary et al. 2022; Fouda et al. 2022b; Puskulluoglu and Michalak. 2022). Few icons are taken from Iconfinder.com for illustration purpose

Table 1.2 Antimicrobial properties of marine plants derived nanoparticles against different pathogenic microorganisms

Source	NPs	Inhibitory activity against	Dose	Inhibition zone/level	Reference
Green seaweed		**Bacteria**			
Caulerpa peltata	ZnONPs	*S. mutans* and *M. luteus*	40 µL	++ and +	Nagarajan and Kuppusamy (2013)
Caulerpa racemosa	AgNPs	*Staphylococcus aureus* and *Proteus mirabilis*	15 µL	12 mm and 14 mm	Kathiraven et al. (2015)
Caulerpa serrulata	AgNPs	*Escherichia coli*, *Shigella* sp., *Staphylococcus aureus*, *Salmonella typhi* and *Pseudomonas aeruginosa*	75 µL	21 mm, ~ 18 mm, ~ 16 mm, ~ 13 mm, and ~ 14 mm	Aboelfetoh et al. (2017)
Caulerpa sertularioides	AgNPs	*Vibrio harveyi*, *V. vulnificus*, and *V. parahaemolyticus*	100 µg/mL	22.96 ± 0.6 mm, 20.14 ± 0.6 mm, & 18.06 ± 0.25 mm	Anjali et al. (2022)
Chaetomorpha antennina	AgNPs	*Vibrio harveyi*	–	20 ± 1 mm & 17.8 ± 2 mm	Thanigaivel et al. (2021)
Chaetomorpha antennina	AgNPs	*Escherichia coli*, *Pseudomonas aeruginosa*, *Klebsiella pneumoniae* and *Proteus mirabilis*	20 µL	1.16 cm, 0.81 ± 0.06 cm, 0.44 ± 0.07 cm, and 0.2 ± 0.06 cm	Roy and Anantharaman (2017)
Chlorodesmis hildebrandtii	AgNPs	*Klebsiella pneumoniae*, *Escherichia coli*, and *Proteus mirabilis*	20 µL	1.5 ± 0.5, 0.63 ± 0.05, and 0.23 ± 0.05 cm	Roy and Anantharaman, (2018a)
Codium tomentosum	AgNPs	*Bacillus subtilis*, *Klebsiella pneumoniae*, and *Salmonella typhi*	150 mg/mL	18.36 ± 1.02 mm, 17.68 ± 0.28 mm, and 20.24 ± 0.84 mm	Murugan et al. (2016b)
Enteromorpha compressa	AgNPs	*Escherichia coli*, *Pseudomonas aeruginosa*, *Klebsiella pneumoniae*, *Staphylococcus aureus*, and *Salmonella paratyphi*	100 µL	12.0 ± 0.41 mm, 11.4 ± 0.82 mm, 11.5 ± 0.62 mm, 11.1 ± 0.47 mm, and 10.5 ± 0.42 mm	Vijayan Sri Ramkumar et al. (2017a, b)

(continued)

1.9 Applications of Marine Flora-Derived Nanoparticles (MFNPs)

Table 1.2 (continued)

Source	NPs	Inhibitory activity against	Dose	Inhibition zone/level	Reference
Enteromorpha flexuosa	AgNPs	B. subtilis, B. pumulis, E. faecalis, S. aureus, S. epidermidis, E. coli, and K. pneumoniae	12.5, 6.25, 50, 25, 6.25, 50, and 50 μg mL^{-1}	18 ± 0.8 mm, 19 ± 1.2 mm, 12 ± 0.9 mm, 14 ± 0.7 mm, 20 ± 1.5 mm, 13 ± 0.9 mm, and 10 ± 0.4 mm	Yousefzadi et al. (2014)
Enteromorpha intestinalis	AgNPs	S. aureus (ATCC 25,923), S. typhi (ATCC 14,028), V. cholera (ATCC 14,730), and E. coli (ATCC 25,922)	25, 12.5, 12.5, and 25 μg/mL MIC	+	Raju et al. (2017)
Halimeda gracilis	AgNPs	Proteus mirabilis and Klebsiella pneumoniae	20 μL	2.33 ± 0.2 cm and 1 ± 0.0 cm	Roy and Anantharaman, (2018b)
Halimeda opuntia	SeNPs	Pseudomonas aeruginosa, Vibrio harveyi and Vibrio parahaemolyticus	7.8 μg	+	(Radhika et al. 2022)
Ulva fasciata	AgNPs	Xanthomonas campestris pv. malvacearum	40.00 ± 5.77 μg/mL MIC	14.00 ± 0.58 mm	Rajesh et al. (2012)
Ulva faciata	AgNPs	S. aureus and E. coli	108 ppm	64.5–98% and 56–96%	El-Rafie et al. (2013)
Ulva fasciata	AgNPs	Escherichia coli ATTCC 8739, Proteus mirabilis ATTCC 9240, Micrococcus leutus and Kocuria varians	100 μg/mL	> 8 mm, > 14 mm, > 8 mm, and > 10 mm	Hamouda et al. (2018)
Ulva fasciata	SeNPs	S. aureus and P. aeruginosa	125 and 250 μg/mL MIC		Shahzamani et al. (2022)

(continued)

Table 1.2 (continued)

Source	NPs	Inhibitory activity against	Dose	Inhibition zone/level	Reference
Ulva fasciata	ZnONPs	*B. subtilus, E. coli, S. aureus,* and *P. aeruginosa*	50–6.25 μg/mL MIC	14.7 ± 0.6 mm to 21.7 ± 0.6 mm	Fouda et al. (2022b)
Ulva flexuosa	FeNPs	*S. epidermidis, B. subtilus,* and *B. pumulis*	30 mg/disk	19 ± 0.4 mm, 18 ± 0.3 mm, and 17 ± 0.4	Mashjoor et al. (2018)
Ulva flexuosa	AgNPs	*P. aeruginosa, E. coli, B. subtilis,* and *S. aureus*	-	24 mm, 15 mm, 18 mm and 13 mm	Dixit et al. (2018)
Ulva lactuca	AgNPs	*Bacillus subtilus, Escherichia coli, Klebsiella* sp., *Pseudomonas aeruginosa, Staphylococcus aureus, Streptococcus faecalis,* and *Neisseria gonorrhoeae*	1 mg	30 mm, 26 mm, 17 mm, 27 mm, 25 mm, 13 mm and 11 mm	Amin (2020)
Ulva lactuca	SeNPs	*Lactobacillus, S. aureus,* and *S. mutans*	100 μL/mL	> 30 mm, > 15 mm, and > 25 mm	Vikneshan et al. (2020)
Ulva lactuca	AgNPs	*E. coli, S. typhi, L. monocytogenes, S. aureus,* and *B. cereus*	30 μL	13.67 ± 1.53 mm, 15.00 ± 1.00 mm, 11.67 ± 2.08, mm 10.33 ± 2.08 mm, and 11.83 ± 0.29 mm	Koçer and Özçimen (2022)
Ulva lactuca	AgNPs	*S. aureus, K. pneumoniae, P. vulgaris,* and *B. subtilis*	75 μL of 15 μg/mL	13.1 ± 0.09 mm, 11.7 ± 0.28 mm, 13.8 ± 0.13 mm, and 11.8 ± 0.05 mm	Gurusamy et al. (2019)
Ulva lactuca	ZnONPs	*B. licheniformis, B. pumilus, E. coli,* and *P. vulgaris*	100 μg/mL	26.3 ± 1.6 mm, 21.2 ± 0.9 mm, 24.0 ± 1.0 mm, and 20.3 ± 0.7 mm	Ishwarya et al. (2018a)
Ulva lactuca	FeNPs	*E. coli, S. aureus, S. typhimurium,* and *P. vulgaris*	20 μL	29 ± 1 mm, 17 ± 2 mm, 26 ± 1 mm, and 22 ± 2 mm	Bensy et al. (2022)
Ulva reticulata	AgNPs	*Bacillus* sp.and *Staphylococcus aureus*	50 μL	26 mm and 25 mm	Bhimba and Devi (2014)

(continued)

1.9 Applications of Marine Flora-Derived Nanoparticles (MFNPs)

Table 1.2 (continued)

Source	NPs	Inhibitory activity against	Dose	Inhibition zone/level	Reference
Ulva rigida	AuNPs	E. coli and S. aureus	128 µg mL^{-1} MLC	17 mm and 12 mm	Algotiml et al. (2022a)
Ulva rigida	AgNPs	B. cereus ATCC11778, S. aureus ATCC25923 and E. coli ATCC25922	64 µg mL^{-1} MLC	16 ± 1, 14 ± 1, and 19 ± 1	Algotiml et al. (2022b)
Urospora sp.	AgNPs	S. aureus, Bacillus subtili, and E. coli	100 µL	23 mm, 20 mm, and 18 mm	Suriya et al. (2012)
Valonopsis pachynema	AgNPs	Micrococcus luteus and Serratia marcescens	20 µL (150 ppm)	9.33 ± 0.33 mm and 8.00 ± 0.58 mm	Selvaraj et al. (2020)
Fungi					
Caulerpa peltata	ZnONPs	C. albicans and A. niger	40 µL	+ + + and + +	Nagarajan and Kuppusamy (2013)
Enteromorpha compressa	AgNPs	Aspergillus flavus, Aspergillus niger, Aspergillus ochraceus, Aspergillus terreus, and Fusarium moniliforme	100 µL	10 ± 0.82, 10.2 ± 0.47, 9.5 ± 0.14, 9.2 ± 0.47, and 9.4 ± 0.84	Vijayan Sri Ramkumar et al. (2017a, b)
Enteromorpha flexuosa	AgNPs	C. albicans and S. cerevisiae	25, and 25 µg mL^{-1}	14 ± 0.8, and 16 ± 0.6	Yousefzadi et al. (2014)
Enteromorpha intestinalis	AgNPs	C. albicans (IFM 40,009), C. tropicalis (IFM 46,521), and C. krusei (IFM 55,058)	25 µg/mL MIC	+	Raju et al. (2017)
Ulva fasciata	Ag/AgCl NPs	C. albicans and C. glabrata	42.98 µg/mL and 21.37 µg/mL IC50	50%	Lashgarian et al. (2021)

(continued)

Table 1.2 (continued)

Source	NPs	Inhibitory activity against	Dose	Inhibition zone/level	Reference
Ulva lactuca	AgNPs	Alternaria alternata, Aspergillus fugigatus, Candida ablicans, Fusarium oxysporum, and Penicillium sp.	1 mg	0.5 mm, 1 mm, 0.5 mm, 1 mm, and 1 mm	Amin (2020)
Ulva lactuca	SeNPs	Candida ablicans	100 µL/mL	> 30 mm	Vikneshan et al. (2020)
Ulva lactuca	AgNPs	Fusarium oxysporum, Botrytis cinerea, and Colletotrichum gloeosporioides	30 µL	54.33 ± 1.53 mm, 71.00 ± 2.00 mm, and 70.33 ± 2.08 mm	Koçer and Özçimen (2022)
Ulva lactuca	AgNPs	Fusarium oxysporum f.sp. vasinfectum and Xanthomonas campestris pv. Malvacearum	80 µg mL^{-1} and 43.33 ± 2.11 µg mL^{-1}	+	Sahayaraj et al. (2018)
Ulva reticulata	AgNPs	Candida albicans, Candida parapsilosis and Aspergillus niger	50 µL	36 mm, 30 mm and 30 mm	Bhimba and Devi (2014)
Ulva rigida	AgNPs	Aspergillus fumigatus	100 µg mL^{-1}	21.3 ± 0.37 mm	El-Kassas and ElKomi (2014)
Ulva rigida	AuNPs	Cryptococcus neoforrmans and C. albicans ATCC 10,231	128 µg mL^{-1} MLC	11 mm and 13 mm	Algotiml et al. (2022a)
Ulva rigida	AuNPs	T. cataneum, and T. mantigrophytes	64 µg mL^{-1} MLC	30 ± 1 mm and 25 mm	Algotiml et al. (2022a)
Ulva rigida	AgNPs	Cryptococcus neoforrmans and C. albicans ATCC 10,231	64 µg mL^{-1} MLC	11 mm, and 13 mm	Algotiml et al. (2022b)
Ulva rigida	AgNPs	T. cataneum, and T. mantigrophytes	32 µg mL^{-1} MLC	30 mm and 40 mm	Algotiml et al. (2022b)
Brown seaweed		Bacteria			

(continued)

1.9 Applications of Marine Flora-Derived Nanoparticles (MFNPs)

Table 1.2 (continued)

Source	NPs	Inhibitory activity against	Dose	Inhibition zone/level	Reference
Bifurcaria bifurcata	CuONPs	Enterobacter cerogenes and Staphylococcus aureus	20 μL	14 mm and 16 mm	Abboud et al. (2014)
Colpomenia sinuosa	AgNPs	S. aureus and E. coli	108 ppm	52–95.6% and 49.5–94%	El-Rafie et al. (2013)
Colpomenia sinuosa	Fe_3O_4NPs	S. typhi and V. cholera	30 mg/mL	11 mm and 15 mm	Salem et al. (2019)
Cystoseira crinita	ZnONPs	Bacillus cereus, Staphylococcus aureus, Escherichia coli, and Salmonella typhi	100 μL (10 mg mL^{-1})	26.0 ± 0.57 mm, 27.3 ± 0.33 mm, 24.3 ± 0.33 mm, and 23.6 ± 0.88 mm	Elrefaey et al. (2022)
Cystoseira crinita	MgONPs	Bacillus subtilis, E. coli, S. aureus, and P. aeuroginosa	50 μg mL^{-1}	13.3 ± 0.5 mm, 13.6 ± 0.5 mm, 17.3 ± 0.5 mm, and 17.6 ± 0.5 mm	Fouda et al. (2022a)
Cystoseira myrica	AgNPs	B. cereus ATCC11778, S. aureus ATCC25923 and E. coli ATCC25922	64 μg mL^{-1} MLC	19 ± 1, 17 ± 1, and 13 ± 1	Algotiml et al. (2022b)
Cystoseira myrica	AuNPs	Escherichia coli and Staphylococcus aureus	5 μL	19 mm and 20.5 mm	Kamal et al. (2022)
Cystoseira trinodis	AuNPs	Escherichia coli and Staphylococcus aureus	5 μL	22 mm and 18 mm	Kamal et al. (2022)
Ecklonia cava	AuNPs	Escherichia coli ATCC 10,536, Bacillus subtils ATCC 6633, Pseudomonas aeruginosa ATCC 27,853 and Staphylococcus aureus ATCC 6538	20 μL	31.8 ± 0.32 mm, 19.7 ± 0.21 mm, 21.3 ± 0.28 mm, and 16.6 ± 0.30 mm	Venkatesan et al. (2014)
Ecklonia cava	AgNPs	E. coli and S. aureus	40 μg	12 ± 1 mm and +	Venkatesan et al. (2016)

(continued)

Table 1.2 (continued)

Source	NPs	Inhibitory activity against	Dose	Inhibition zone/level	Reference
Fucus evanescens	AgNPs	*E. coli*	10 µg/disk	3–3.5 mm	Yugay et al. (2020)
Fucus vesiculosus		*P. aeruginosa* PAO1 KCTC 1637	512 µg/mL MIC	Positive	Khan et al. (2019)
Halopteris scoparia	AgNPs	*E. coli, S. typhi, L. monocytogenes, S. aureus,* and *B. cereus*	30 µL	14.67 ± 0.58 mm, 14.67 ± 3.05 mm, 10.33 ± 0.58 mm, 10.67 ± 0.58 mm, and 10.00 ± 1.73 mm	Koçer and Özçimen (2022)
Laminaria ochroleuca	AgNPs	*S. aureus* and *E. coli*	80 µL	21 mm and 15 mm	Kaidi et al. (2022)
Padina sp.	AgNPs	*Staphylococcus aureus* and *Pseudomonas aeruginosa*	1.00 mg/mL	15.17 ± 0.58 mm and 13.33 ± 0.76 mm	Bhuyar et al. (2020)
Padina boryana	PdNPs	*Staphylococcus aureus, Escherichia fergusonii, Acinetobacter pittii, Pseudomonas aeruginosa, Aeromonas enteropelogenes,* and *Proteus mirabilis*	100 µL of 1 mg/mL	18.3 ± 1.24 mm, 20.0 ± 0.81 mm, 23.0 ± 0.8 mm, 21.3 ± 0.47 mm, 19.3 ± 0.5 mm and 23.0 ± 1.6 mm	Sonbol et al. (2021)
Padina gymnospora	AgNPs	*Bacillus cereus* and *Escherichia coli*	–	13.06 ± 0.40 mm and 9.5 ± 0.2 mm	Shiny et al. (2013a, b)
Padina gymnospora	PtNPs	*E. coli, K. pneumoniae, L. lactis, S. typhi, S. aureus, S. mutans,* and *S. pneumoniae*	75, 75, 75, 100, 100, 75, and 100 µg/mL MIC	3 ± 0.05, 2.5 ± 0.16, 2.4 ± 0.08, 2.1 ± 0.06, 2.2 ± 0.05, 2 ± 0.05, and 2 ± 0.13	Sri Ramkumar et al. (2017a, b)
Padina pavonica	AgNPs	*Xanthomonas campestris* pv. *malvacearum*	100 µL	10.33 ± 0.33 mm	Sahayaraj et al. (2012)
Padina pavonica	AuNPs	*Escherichia coli*	5 mg/mL	15 mm	Isaac and Renitta (2015)

(continued)

1.9 Applications of Marine Flora-Derived Nanoparticles (MFNPs)

Table 1.2 (continued)

Source	NPs	Inhibitory activity against	Dose	Inhibition zone/level	Reference
Padina pavonica	AgNPs	*Escherichia coli* and *Staphylococcus aureus*	30 μL	7.50 ± 0.09 mm and 8.60 ± 0.09 mm	Sudha and Balasundaram, (2018)
Padina tetrastromatica	AgNPs	*Bacillus* sp., *Klebsiella planticola*, *Bacillus subtils*, and *Pseudomonas* sp.			Rajeshkumar et al. (2012a)
Padina tetrastromatica	AgNPs	*Staphyloccocus aureus* and *Pseudomonas aeruginosa*	40 μL	4 mm and 6 mm	Sangeetha et al. (2012)
Padina tetrastomatica	AgNPs	*P. mirabilis*, *B. cereus*, *E. coli*, and *S. aureus*	-	~ 14 mm, < 15 mm, ~ 15 mm, and < 14 mm	Shiny et al. (2013a, b)
Padina tetrastromatica	AgNPs	*Bacillus subtilis*, *Bacillus* sp., *Serratia nematodiphila*, *Klebsiella planticola*, *Klebsiella Pneumoniae* and *Streptococcus* sp.	25 μL of 100 nM	12–15 mm	Rajeshkumar (2017)
Saccharina cichorioides	AgNPs	*E. coli*	10 μg/disk	1.5–3 mm	Yugay et al. (2020)
Saccharina japonica	AgNPs	*E. coli*, *S. typhimurium*, *L. monocytogenes*, *B. cereus*, and *S. aureus*	100 μL	20 ± 0.70 mm, 16 ± 0.10 mm, 16 ± 0.25 mm, 14 ± 0.25 mm and 22 ± 0.05 mm	Sivagnanam et al. (2017)
Sargassum cinereum	AgNPs	*Staphylococcus aureus*	2.5 μL MIC		Mohandass et al. (2013)
Sargassum cinereum	AgNPs	*Enterobacter aerogenes*, *Salmonella typii* and *Proteus vulgaris*	100 μg/disk		Mohandass et al. (2013)

(continued)

Table 1.2 (continued)

Source	NPs	Inhibitory activity against	Dose	Inhibition zone/level	Reference
Sargassum dentifolium	AgNPs	Salmonella typhimurium (ATCC14028), Enterobacter aerogenes (ATCC13048), Pseudomonas aeruginosa (ATCC278223), Escherichia coli (ATCC 25922) and methicillin-resistant Staphylococcus aureus (MRSA, ATCC43300)	10^{-4}–10^{5}/mL MIC	+	Saber et al. (2017)
Sargassum ilicifolium	AgNPs	E. coli, K. pneumoniae, S. typhi, S. aureus, and V. cholerae	100 nM	18.2, 16.2, 17.1, 16.8, and 17.3	Kumar et al. (2012a)
Sargassum ilicifolium	AgNPs	Escherichia coli, Pseudomonas aeruginosa, Proteus mirabilis and Klebsiella pneumoniae	20 μL	0.83 ± 0.05 cm, 0.53 ± 0.05 cm, 0.46 ± 0.05 cm, and 0.4 ± 0.1 cm	Suparna Roy and Anantharaman, (2018a, b, c, d)
Sargassum incisifolium	AgNPs	A. baumannii, K. pneumonia, E. faicalis, and S. aureus	0.26 mM	12.8 ± 0.20 mm, 7.5 ± 0.26 mm, 7.0 ± 0.19 mm, and 12.1 ± 0.23 mm	Mmola et al. (2016)
Sargassum incisifolium	AuNPs	K. pneumonia, and E. faicalis, and S. aureus,	0.21 mM	2.5 ± 0.5 mm, 1.5 ± 0 mm, and 4.0 ± 0.25	Mmola et al. (2016)
Sargassum longifolium	CuONPs	Serratia marcescens, Vibrio parahemolyticus, and Aeromonas hydrophila	25 μL, 50 μL, & 100 μL	14 ± 0.34 mm, 16 ± 0.18 mm, and 17 ± 1.12 mm	Rajeshkumar et al. (2021)
Sargassum muticum	AgNPs	Bacillus subtilis, Klebsiella pneumoniae, and Salmonella typhi	150 ppm	13.2 ± 0.6 mm, 14.4 ± 0.6 mm, and 15.4 ± 0.4 mm	Madhiyazhagan et al. (2015)
Sargassum muticum	AgNPs	Staphylococcus aureus, Staphylococcus saprophyticus, and Enterococcus faecalis	19.75 (39.50), 39.50 (*), and 39.50 (*) MIC(MBC) μg/mL	+	González-Ballesteros et al. (2020)

(continued)

1.9 Applications of Marine Flora-Derived Nanoparticles (MFNPs)

Table 1.2 (continued)

Source	NPs	Inhibitory activity against	Dose	Inhibition zone/level	Reference
Sargassum muticum	AuNPs	Staphylococcus aureus, Staphylococcus saprophyticus, and Enterococcus faecalis	3.38 (6.75), 6.75 (*), and 13.50 (*) MIC(MBC) μg/mL	+	González-Ballesteros et al. (2020)
Sargassum myriocystum	ZnONPs	S. mutans and M. luteus	40 μL	++ and +	Nagarajan and Kuppusamy (2013)
Sargassum myriocystum	AuNPs	Pesudomonas aeroginesa, Klebsiella oxyctoca, Enterobacter fecalis, K. pneumonia, V.cholerae, S.typi, S. paratype, V. parahaemolyticus, and P. vulgaris	100 μL	8 mm, 7 mm, 11 mm, 6 mm, 8 mm, 6 mm, 8 mm, 9 mm, and 8 mm	Ismail et al. (2018)
Sargassum myriocystum	AgNPs	Bacillus subtilis, E. coli, Pseudomonas sp. and Klebshella planticola	–	Positive	Kumar and Rajeshkumar (2017)
Sargassum myriocystum	AgNPs	S. epidermidis, S. aureus, E. coli, K. pneumoniae, P. aeruginosa, and Pseudomonas valgaris	10 μL of 50 μg/mL	18.6 ± 1.764 mm, 17.3 ± 1.453 mm, 15.0 ± 1.720 mm, 15.0 ± 1.155 mm, 17.6 ± 0.881 mm, and 12.0 ± 1.732 mm	Balaraman et al. (2020)
Sargassum plagiophyllum	AgClNPs	E. coli	20 μg/mL	Positive	Dhas et al. (2014a, b)
Sargassum polycystum	AgNPs	Pseudomonas aeruginosa, Klebsiella pneumoniae, Escherichia coil, and Staphylococcus aureus	50 μL	23 mm, 22 mm, 28 mm, and 26 mm	Thangaraju et al. (2012)
Sargassum polycystum	CuONPs	Pseudomonas aeruginosa and Shigella dysenteriae	50 μg/mL	15 ± 0.5 mm and 6 ± 0.5 mm	Ramaswamy et al. (2016)

(continued)

Table 1.2 (continued)

Source	NPs	Inhibitory activity against	Dose	Inhibition zone/level	Reference
Sargassum polycystum	AgNPs	*E. coli* and *Morganella morganii*	100 μL/mL	5 ± 0.5 mm and 8 ± 0.3 mm	Asha et al. (2015)
Sargassum polycystum	AgNPs	*Staphylococcus aureus*, *Micrococcus luteus*, *Pseudomonas fluorescens*, *Serratia marcescens*, *Klebsiella pneumoniae*, *Bacillus subtilis*, *Escherichia coli*, *Staphylococcus epidermidis* and *Vibrio cholera*	50 μL of 100 μg/mL	36 mm, 35 mm, 25 mm, 18 mm, 18 mm, 18 mm, 17 mm, 17 mm, and 15 mm	Thiurunavukkarau et al. (2022)
Sargassum polycystum	AuNPs	*E. coli*, *S. typhimurium*, *P. aeruginosa*, *S. dysenteriea*, and *V. cholera*	30 μg/mL	~ 11 mm, ~ 9 mm, ~ 8.5 mm, and ~ 11 mm	Sivaraj et al. (2015)
Sargassum polyphyllum	AgNPs	*S. aureus*, *P. aeruginosa*, *B. subtilis*, and *E. coli*	20 μL	13 mm, 17 mm, 14 mm, and 18 mm	Arunkumar et al. (2014)
Sargassum swartzii	ZnONPs	*V. parahaemolyticus*	25 μg mL^{-1} MIC	15 ± 2.0 mm,	Vinu et al. (2021)
Sargassum swartzii	CuONPs	*V. parahaemolyticus*	25 μg mL^{-1} MIC	13 ± 1.20 mm	Vinu et al. (2021)
Sargassum swartzii	SeNPs	*V. parahaemolyticus*	10 μg mL^{-1} MIC	13 ± 1.65 mm	Vinu et al. (2021)
Sargassum swartzii	AgNPs	*Escherichia coli* ATCC 25,922	20 mg/mL	Positive	Dhas et al. (2021)
Sargassum tenerrimum	AgNPs	*Vibrio cholerae*	30 μL	18 mm	Kumar et al. (2012a, b, c)

(continued)

1.9 Applications of Marine Flora-Derived Nanoparticles (MFNPs) 63

Table 1.2 (continued)

Source	NPs	Inhibitory activity against	Dose	Inhibition zone/level	Reference
Sargassum wightii	AgNPs	*Staphylococcus aureus, Bacillus rhizoids, Escherisia coli,* and *Pseudomonas aeruginosa*	50 μL	13 mm, 14 mm, 14 mm, and 15 mm	Govindaraju et al. (2009)
Sargassum wightii	AgNPs	*E. coli,* followed by *S. flexneri, S. typhimurium, E. cereus, B. subtilis* and *E. faecalis*	–	12–16 mm (range)	Deepak et al. (2018)
Sargassum wightii	AgNPs	*P. aeruginosa, V. cholera, K. pneumonia, S. aureus, E. coli, S. pneumoniae* and *S. typhi*	30 μL	5 mm, 5 mm, 7 mm, 12 mm, 3 mm, 1 mm, and 7 mm	Shanmugam et al. (2014)
Sargassum wightii	AgNPs	*Escherichia coii, Staphylococcus aureus, Pseudomonas aeruginosa* and *Klebsiella pneumoniae*	100% of dry AgNPs powder	11 mm, 9 mm, 9 mm, and 9 mm	Sunitha et al. (2015)
Sargassum wightii	AgNPs	*Micrococcus luteus* and *Serratia marcescens*	20 μL (150 ppm)	23.33 ± 2.90 mm and 14.33 ± 3.05 mm	Selvaraj et al. (2020)
Sargassum wightii	MgONPs	MRSA 56 and *P. aeruoginosa*	256 μg/mL MIC		Pugazhendhi et al. (2019)
Sargassum wightii	AgNPs	*Enterococcus* sp. and *Staphylococcus* sp.	100 μL	5 mm and 5 mm	Suganya et al. (2020)
Sargassum wightii	AgNPs	*Bacillus cereus, Bacillus anthracis, Staphylococcus aureus* and *Vibrio alginoyticus*	130 μg/mL MIC	10 mm, 8 mm, 10 mm and 9 mm	Thirumalairaj et al. (2014)

(continued)

Table 1.2 (continued)

Source	NPs	Inhibitory activity against	Dose	Inhibition zone/level	Reference
Sargassum wightii	AgNPs	*Micrococcus luteus, Staphyloccocus aureus, Pseudomonas fluorescens, Klebsiella pneumoniae, Escherichia coli, Staphylococcus epidermidis, Bacillus subtilis, Serratia marcescens*	50 µL of 100 µg/mL	30 mm, 27 mm, 20 mm, 16 mm, 15 mm, 14 mm, 12 mm, and 12 mm	Thirunavukkarau et al. (2022)
Sargassum wightii	ZrO_2NPs	*Bacillus subtilis, Escherichia coli,* and *Salmonella typhi*	15 µg	21 mm, 19 mm, and 19 mm	Kumaresan et al. (2018)
Spatoglossum asperum	AgNPs	*Klebsiella pneumoniae*	150 µg/mL	18 ± 0.28 mm	Ravichandran et al. (2018)
Stoechospermum marginatum	AuNPs	*P. aeruginosa, K. oxytoca, E. faecalis, K. pneumoniae, V. cholerae, S. typhi, S. paratyphii, V. parahaemolyticus,* and *P. vulgaris*	100 µL	8 mm, 7 mm, 11 mm, 6 mm, 8 mm, 6 mm, 8 mm, 9 mm, 8 mm	Rajathi et al. (2012)
Turbinaria conoides	AgNPs	*B. subtilis* and *Klebsiella planticola*	300 µL	10.33 ± 0.88 mm and 20.67 ± 0.66 mm	Rajeshkumar et al. (2012b)
Turbinaria conoides	AgNPs	*Salmonella* sp., *E. coli, S. liquefaciens,* and *A. hydrophila*	20 µL mL^{-1} MIC, 20 µL mL^{-1} MIC, 40 µL mL^{-1} MIC, and 40 µL mL^{-1} MIC,	14.5 ± 0.41 mm, 15.6 ± 0.42 mm, 13.2 ± 0.62 mm, and 12.3 ± 0.47 mm	Vijayan et al. (2014)
Turbinaria conoides	AuNPs	*Bacillus subtilis, Klebsiella planticola* and *Streptoccocus* sp.	90 µL	12 mm, 13 mm, and 14 mm	Rajeshkumar et al. (2013a, b)
Turbinaria ornata	AgNPs	*S. aureus, B. circulans, E.coli, E. faecalis,* and *P. aeruginosa*	100 µL	~ 13 mm, 12 mm, 13 mm, 12 mm, and 12 mm	Anuluxan et al. (2022)
		Fungi			
					(continued)

1.9 Applications of Marine Flora-Derived Nanoparticles (MFNPs)

Table 1.2 (continued)

Source	NPs	Inhibitory activity against	Dose	Inhibition zone/level	Reference
Colpomenia sinuosa	Fe_3O_4NPs	*Aspergillus fumigatus* and *Fusarium moroliforme*	30 mg/mL	11 mm and 15 mm	Salem et al. (2019)
Cystoseira crinita	ZnONPs	*Candida albicans* and *Aspergillus niger*	100 μL (10 mg mL^{-1})	22.3 ± 0.33 mm and 31.0 ± 0.57 mm	Elrefaey et al. (2022)
Cystoseira crinita	MgONPs	*C. albicans*	50 μg mL^{-1}	16.6 ± 0.5 mm	Fouda et al. (2022a)
Cystoseira myrica	AgNPs	*Cryptococcus neoformans* and *C. albicans* ATCC 10,231	64 μg mL^{-1} MLC	13 ± 0.6 mm and 15 ± 1 mm	Algotiml et al. (2022b)
Cystoseira myrica	AgNPs	*T. cataneum*, and *T. mantigrophytes*	32 μg mL^{-1} MLC	30 mm and 30 mm	Algotiml et al. (2022b)
Cystoseira trinodis	AuNPs	*Aspergillus niger, Alternaria alternate*, and *Candida albicans*	5 μL	18 mm, 17 mm and 16 mm	Kamal et al. (2022)
Dictyota bartayresiana	AuNPs	*Fusarium dimerum*	50 μL	12 mm	Varun et al. (2014)
Ecklonia cava	AuNPs	*Aspergillus niger* ATCC 1015, *A. brasiliensis* ATCC 16404, *A. fumigates* ATCC 1022, and *Candida albicans* ATCC 10231	20 μL	24.6 ± 0.23 mm, 19.3 ± 0.26 mm, 21.5 ± 0.25 mm, and 23.3 ± 0.25 mm	Venkatesan et al. (2014)
Halopteris scoparia	AgNPs	*Fusarium oxysporum, Botrytis cinerea*, and *Colletotrichum gloeosporioides*	30 μL	49.00 ± 1.73 mm, 70.33 ± 2.08 mm, and 68.67 ± 2.89 mm	Koçer and Özçimen (2022)
Lobophora variegata	AgNPs	*Candida albicans, Candida tropicalis, Trichophyton mentagrophytes*, and *Aspergillus flavus*	100 μg/mL	16.54 ± 0.002, 14.78 ± 0.002, 16.25 ± 0.002, and 12.71 ± 0.002	Sathyaseelan et al. (2015)

(continued)

Table 1.2 (continued)

Source	NPs	Inhibitory activity against	Dose	Inhibition zone/level	Reference
Padina australis	AgNPs	*Rhizoctonia solani* and *Xanthomanas oryzae*	100 μL	16 ± 1.6 mm and 15 ± 1.3 mm	Kailasam et al. (2023)
Padina pavonica	AgNPs	*Fusarium oxysporum* f.sp. *vasinfectum*	100 μL	12.33 ± 0.33 mm	Sahayaraj et al. (2012)
Padina tetrastromatica	AgNPs	*Fusarium sp, Aspergillus niger, Candida albicans, Aspergillus fumigatus,* and *Aspergillus sp.*	150 μL	20.03 ± 0.033 mm, and 18.13 ± 0.089 mm, 12.20 ± 0.152 mm, 12.20 ± 0.100 mm, and 10.17 ± 0.167 mm	Rajeshkumar et al. (2017b)
Saccharina japonica	AgNPs	*C. albicans*	100 μL	15 ± 0.10 mm	Sivagnanam et al. (2017)
Sargassum angustifolium	SeNPs	*Vibrio harveyi*	200 μg/mL MIC	+	Mansouri-Tehrani et al. (2021)
Sargassum incisifolium	AgNPs	*C. albicans*	0.26 mM	17.8 ± 0.65 mm	Mmola et al. (2016)
Sargassum muticum	AgNPs	*Bacillus subtilis, Escherichia coli, Klebsiella pneumoniae* and *Salmonella typhi*	-	11.25 mm, 13.35 mm, 14.24 mm, and 12.23 mm	Trivedi et al. (2021)
Sargassum myriocystum	ZnONPs	*C. albicans* and *A. niger*	40 μL	+++ and ++	Nagarajan and Kuppusamy (2013)
Sargassum polycystum	CuONPs	*Aspergillus niger* and *Aspergillus oryzae*	50 μg/mL	20 ± 0.5 mm and 12 ± 0.5 mm	Ramaswamy et al. (2016)
Sargassum polycystum	AgNPs	*Candida albicans*	50 μL of 100 μg/mL	17 mm	Thiurunavukkarau et al. (2022)
Sargassum wightii	AgNPs	*Candida albicans*	50 μL of 100 μg/mL	15 mm	Thiurunavukkarau et al. (2022)

(continued)

1.9 Applications of Marine Flora-Derived Nanoparticles (MFNPs)

Table 1.2 (continued)

Source	NPs	Inhibitory activity against	Dose	Inhibition zone/level	Reference
Spatoglossum asperum	AgNPs	Candida albicans, Candida tropicalis, Trichophyton mentagrophytes, and Aspergillus flavus	100 μg/mL	20.67 ± 0.88 mm, 17.67 ± 0.33 mm, 17.33 ± 0.88 mm, and 12.67 ± 0.33 mm	Subbiah et al. (2019)
Red seaweed		**Bacteria**			
Marine red algae	Co_3O_4NPs	S. aureus, B. subtilis, E. coli, and P. aeruginosa	25.0 ± 7.3 μg/mL MIC	21.1 ± 7.1 mm, 23.6 ± 6.9 mm, and 20.8 ± 5.8 mm	Ajarem et al. (2022)
Acanthophora spicifera	AgNPs	E. coli, S. typhii, S. flexneri, S. aureus, and V. cholerae	100 nM	14 ± 2 mm, 15 mm, 15 mm, 11.75 ± 0.5 mm, and 12.75 ± 2.94 mm	Kumar et al. (2012b)
Acanthophora spicifera	AgNPs	Escherichia coli, Salmonella sp., Bacillus subtilis & Staphylococcus aureus	200 μL	23.5 mm, 19.5 mm, 19.5 mm, & 18.75 mm	Ibraheem et al. (2016)
Acanthophora spicifera	AuNPs	Vibrio harveyi and Staphylococcus aureus	100 μg/mL	22 ± 0.3 mm and 18.7 ± 0.5 mm	Babu et al. (2020)
Acanthophora spicifera	AgNPs	Micrococcus luteus, Staphylococcus aureus, Pseudomonas fluorescens, Escherichia coli, Serratia marcescens, Klebsiella pneumoniae, Bacillus subtilis, Staphylococcus epidermidis, and Vibrio cholere	50 μL of 100 μg/mL	36 mm, 36 mm, 24 mm, 18 mm, 16 mm, 17 mm, 17 mm and 14 mm	Thirunavukkarau et al. (2022)
Amphiroa anceps	AgNPs	Escherichia coli & Klebsiella pneumoniae	–	0.95 ± 0.057 cm & 0.63 ± 0.05 cm	Roy and Anantharaman, (2018c)

(continued)

Table 1.2 (continued)

Source	NPs	Inhibitory activity against	Dose	Inhibition zone/level	Reference
Amphiroa fragilissima	AgNPs	Bacillus subtilis, Bacillus cereus, Staphylococcus aureus, Escherichia coli, Salmonella typhi, Shigella dysentriae, and Pseudomonas aeruginosa	100 µL	19 mm, 12 mm, 20 mm, 13 mm, 15 mm, 10 mm, and 13 mm	Ramalingam et al. (2018)
Amphiroa rigida	AgNPs	Staphylococcus aureus and Pseudomonas aeruginosa	3.125 µg/mL MIC and 6.25 µg/mL MIC	21 ± 0.2 mm & 15 ± 0.2 mm	Gopu et al. (2021)
Champia parvula	AgNPs	Streptococcus mutans, Staphylococcus aureus, and Enterococcus faecalis	100 µL/mL	23 mm, 21 mm, and 20 mm	Viswanathan et al. (2023)
Corallina elongata	AgNPs	Escherichia coli ATCC 8739, Micrococcus leutus, and Kocuria varians	0.1 mM	11 mm, 11 mm, and 8 mm	Hamouda et al. (2019)
Galaxaura elongata	AuNPs	E. coli, K. pneumoniae, P. aeruginosa, S. aureus, and S. aureus (MRSA)		13.5–17 mm, 13–17 mm, 9–13 mm, 9–13 mm, and 7.5–16 mm	Abdel-Raouf et al. (2017)
Gelidiella acerosa	AuNPs	Bacillus subtilis, E. coli, Serratia marcescens and Klebsiella pneumonia	20 µL	14–18 mm (range)	Senthilkumar et al. (2019)
Gelidiella acerosa	AuNPs	Staphylococcus aureus			Subbulakshmi et al. (2022)
Gelidiella acerosa	AgNPs	Pseudomonas aeruginosa, Escherichia coli, Staphylococcus aureus, and Bacillus subtilis	100 µL	0.7 ± 0.2 mm, 0.6 ± 0.2 mm, and 1.6 ± 0.3 mm	Thiruchelvi et al. (2021)

(continued)

1.9 Applications of Marine Flora-Derived Nanoparticles (MFNPs)

Table 1.2 (continued)

Source	NPs	Inhibitory activity against	Dose	Inhibition zone/level	Reference
Gelidium amansii	AgNPs	*A. hydrophila, E. coli, P. aeruginosa,* and *V. parahaemolyticus*	100 μg	90%	Pugazhendhi et al. (2018)
Gelidium amansii	AgNPs	*Escherichia coli* ATCC 8739, *Micrococcus luteus,* and *Kocuria varians*	0.1 mM	12 mm, 8 mm, and 10 mm	Hamouda et al. (2019)
Gelidium amansii	AuNPs	*S. aureus* and *E. coli*	300 ppm	+	Kumar et al. (2017a, b)
Gelidium corneum	AgNPs	*Escherichia coli* 25,922	0.51 μg/mL (MBC) and 0.26 μg/mL (MIC)	–	Öztürk et al. (2020)
Gracilaria sp.	AuNPs	*Staphylococcus aureus*	100% and 50%	12 mm and 8 mm	Ramakritinan et al. (2013)
Gracilaria birdiae	AgNPs	*Escherichia coli* and *Staphylococcus aureus*	34.3 μg/mL MIC & 81.2 μg/mL MIC	–	Aragao et al. (2019)
Gracilaria corticata	AgNPs	*B. subtilis, S. aureus, P. aeruginosa,* and *P. vulgaris*	100 μg/mL	15 mm, 12 mm, 12 mm, and 10 mm	Parthasarathy et al. (2021)
Gracilaria corticata	AuNPs	*E. coli, Enterobacter aerogenes, Staphylococcus aureus,* and *Enterococcus faecalis*		24 mm, 21 mm, 19 mm, and 14 mm	Naveena and Prakash (2013)
Gracilaria corticata	AgNPs	*E. coli, P. aeruginosa,* and *V. cholerae*	1 mg/mL	11 mm, 12 mm, and 11 mm	Aravindan et al. (2014)
Gracilaria crassa	AgNPs	*Escherichia coli, Proteus mirabilis, Bacillus subtilis* and *Pseudomonas aeruginosa*	40 μg/mL	~ 42 mm, ~ 38 mm, ~ 22 mm, and ~ 23 mm	Lavakumar et al. (2015)

(continued)

Table 1.2 (continued)

Source	NPs	Inhibitory activity against	Dose	Inhibition zone/level	Reference
Gracilaria dura	AgNPs	*Bacillus pumilus* HQ318731	Agar/silver nanocomposite films (1.5 g agar + 100 mL of silver nanoparticles solution) as disks	25 mm	Shukla et al. (2012)
Gracilaria foliifera	AgNPs	*B. cereus* ATCC11778, *S. aureus* ATCC25923 and *E. coli* ATCC25922	64 μg mL^{-1} MLC	14 ± 1.5, 18 ± 0.6, and 14 ± 1	Algotiml et al. (2022b)
Gracilaria gracilis	AgNPs	*S. typhimorium*, *E. coli*, *K. pneumonia*	29 μg /mL		Kochesfehani et al. (2021)
Gracilaria parvispora	AgNPs	*Staphylococcus aureus* and *Pseudomonas aeruginosa*	50 μL	+	Hussein et al. (2017)
Halymenia porphyriformis	AgNPs	*Staphylococcus aureus* MT416445, *Streptococcus viridans* MT416448, *Lactobacillus acidophilus* MT416447 and *Lactobacillus brevis* MT416446	200 μg	4.6 ± 2.18 mm, 8.9 ± 0.95 mm, 11.6 ± 1.66 mm, and 10 ± 2.88 mm	Khan et al. (2022a, b)
Halymenia porphyroides	AgNPs	*Staphylococcus aureus*, *Salmonella typhi*, *Escherichia coli*, *Klebsiella pneumonia*, and *Proteus vulgaris*	30 μg/mL	20 mm, 22 mm, 20 mm, 22 mm, and 20 mm	Kiran and Murugesan (2014b)
Halymenia pseudofloresti	AuNPs	*Staphylococcus aureus*, *Lactobacillus*, and *Pseudomonas aeruginosa*	100 μg/mL	24 mm, 23 mm, and 22 mm	Palaniyandi et al. (2023)

(continued)

1.9 Applications of Marine Flora-Derived Nanoparticles (MFNPs)

Table 1.2 (continued)

Source	NPs	Inhibitory activity against	Dose	Inhibition zone/level	Reference
Halymenia venusta	AuNPs	*Staphylococcus aureus, Pseudomonas aeruginosa,* and *Klebsiella pneumoniae*	100 µL	+	Baskar et al. (2023)
Hypnea musciformis	AgNPs	*Klebsiella pneumonia, Bacillus subtilis, Staphylococcus aureus, Pseudomonas aeruginosa,* and *Escherichia coli*	80 µL	15 mm, 22 mm, 24 mm, 10 mm, and 26 mm	Devi and Bhimba (2014)
Hypnea musciformis	AgNPs	*X. campestris* and *R. solanacearum*	150 µg/mL	24.8 ± 0.1 mm and 20.3 ± 0.6 mm	Vadlapudi and Amanchy (2017)
Hypnea valentiae	ZnONPs	*S. mutans* and *M. luteus*	40 µL	++ and +	Nagarajan and Kuppusamy (2013)
Hypnea valentiae	AgNPs	*P. aeruginosa, E. faecalis, S. aureus,* and *S. mutans*	50 µg/mL, 100 µg/mL, 25 µg/mL, and 12.5 µg/mL MIC	26 mm, 20 mm, 16 mm and 18 mm	Viswanathan et al. (2022)
Jania rubens	AgNPs	*S. aureus* and *E. coli*	108 ppm	59–96.4% and 50–92.5%	El-Rafie et al. (2013)
Jania rubens	AgNPs	*Salmonella typhimurium* (ATCC14028), *Enterobacter aerogenes* (ATCC13048), *Pseudomonas aeruginosa* (ATCC278223), *Escherichia coli* (ATCC 25,922) and methicillin-resistant *Staphylococcus aureus* (MRSA, ATCC43300)	10^4–10^5/mL MIC	+	Saber et al. (2017)
Kappaphycus	ZnONPs	MRSA	100 µg/mL	15.5 mm	Vijayakumar et al. (2020)

(continued)

Table 1.2 (continued)

Source	NPs	Inhibitory activity against	Dose	Inhibition zone/level	Reference
Kappaphycus alvarezii	SeNPs	Pseudomonas aeruginosa, Vibrio harveyi and Vibrio parahaemolyticus	7.8 µg	+	Radhika et al. (2022)
Kappaphycus alvarezii	AgNPs	Escherichia coli			Khan et al. (2022a, b)
Laurencia papillosa	AgNPs	Bacillus subtilis, Staphylococcus aureus, Streptococcus pneumoniae, Escherichia coli, Klebsiella pneumoniae, and Pseudomonas aeruginosa	150 µL	21 ± 0.5 mm, 19 ± 0.3 mm, 18 ± 0.2 mm, 17 ± 0.2 mm, 18 ± 0.5 mm, and 16 ± 0.3 mm	Omar et al. (2017)
Portieria hornemannii	AgNPs	Vibrio parahaemolyticus, Vibrio vulnificus, Vibrio syhim and Vibrio anguillarum			Fatima et al. (2020)
Portieria hornemannii	AgNPs	E. coli, Staphyloccocus aureus, Klebsiella spp., Proteus spp. and Pseudomonas spp.	80 µL	19 mm, 20 mm, 16 mm, 15 mm, and 9 mm	Sabatini and Anchana Devi (2017)
Porphyra vietnamensis	AgNPs	E. coli and S. aureus	5 µg/mL and 15 µg/mL	100% and 60%	Venkatpurwar and Pokharkar (2011)
Portieria hornemannii	AgNPs	E. coli, P. aeruginosa, V. cholerae and S. aureus	1 mg/mL	11 mm, 13 mm, 13 mm, and 18 mm	Aravindan et al. (2014)
Pterocladiella capillacea	AgNPs	S. aureus and E. coli	108 ppm	63–98% and 52–95.7%	El-Rafie et al. (2013)
Pterocladiella capillacea	AgNPs	Bacillus subtillus and S. aureus	20 mg/mL	25.1 ± 0.25 mm and 18.1 ± 0.22 mm	Kassas and Attia (2014)

(continued)

1.9 Applications of Marine Flora-Derived Nanoparticles (MFNPs)

Table 1.2 (continued)

Source	NPs	Inhibitory activity against	Dose	Inhibition zone/level	Reference
Pterocladia capillacea	Fe$_3$O$_4$NPs	V. cholera	30 mg/mL	9 mm	Salem et al. (2019)
Pyropia yezoensis	AgNPs	Pseudomonas aeruginosa	200 and 400 µg/mL		Ulagesan et al. (2021)
Rhodymenia palmata	AgNPs	Bacillus subtilis	2 µg	7 mm	Murugammal and Flora (2017)
Solieria robusta	AgNPs	Staphylococcus aureus MT416445, Streptococcus viridans MT416448, Lactobacillus acidophilus MT416447 and Lactobacillus brevis MT416446	200 µg	5 ± 1.73 mm, 4.3 ± 1.35 mm, 2.1 ± 0.45 mm, and 6.9 ± 0.98 mm	Khan et al. (2022a, b)
Spyridia filamentosa	AgNPs	Klebsiella sp. and Staphylococcus sp.	20 mM	63.4% and 44.6%	Valarmathi et al. (2020)
Spyridia fusiformis	AgNPs	K. pneumoniae and S. aureus	100 µL	26 mm and 24 mm	Murugesan et al. (2017)
		Fungi			
Acanthophora spicifera	AgNPs	Candida albicans	200 µL	20 mm	Ibraheem et al. (2016)
Acanthophora spicifera	AgNPs	Candida albicans	50 µL of 100 µg/mL	20 mm	Thirunavukkarau et al. (2022)
Champia parvula	AgNPs	Candida albicans	100 µL/mL	20 mm	Viswanathan et al. (2023)

(continued)

Table 1.2 (continued)

Source	NPs	Inhibitory activity against	Dose	Inhibition zone/level	Reference
Gelidiella acerosa	AgNPs	Humicola insolens MTCC 4520, Fusarium dimerum MTCC 6583, Mucor indicus MTCC 3318 and Trichoderma reesei MTCC 3929	50 µL	12.2 mm, 13.15 mm, 22.3 mm, and 17.2 mm	Vivek et al. (2011)
Gelidium corneum	AgNPs	Candida albicans 14.053	2.04 µg/mL (MBC) and 0.51 µg/mL (MIC)	–	Öztürk et al. (2020)
Gracilaria corticata	AgNPs	Candida albicans NCIM 3074 and C. glabrata NCIM 3226	30 µL	12 mm and 11 mm	Kumar et al. (2013b)
Gracilaria foliifera	AgNPs	Cryptococcus neoformans, C. albicans ATCC 10,231, T. cataneum, and T. mantigrophytes	64 µg mL^{-1} MLC	14 ± 0.6 mm, 13 mm, 18 mm, and 23 mm	Algotiml et al. (2022b)
Halymenia porphyroides	AgNPs	Microsporum nanum, Rhizopus microspores, and Trichophyton mentagrophytes	30 µg/mL	2 ± 0.002 mm, 4 ± 0.002 mm and 2 ± 0.001 mm	Manam and Subbaiah (2020)
Hypnea musciformis	AgNPs	Aspergillus niger Candida albicans Candida parasilopsis	80 µL	27 mm, 26 mm, and 21 mm	Devi and Bhimba (2014)
Hypnea musciformis	AuNPs	Aspergillus niger and Mucor sp.	Positive inhibitory activity	–	Murugesan et al. (2015)
Hypnea valentiae	ZnONPs	C. albicans and A. niger	40 µL	+++ and ++	Nagarajan and Kuppusamy (2013)
Laurencia papillosa	AgNPs	Aspergillus flavus, Aspergillus fumigatus, and Aspergillus niger	150 µL	15 ± 0.5 mm, 13 ± 0.3 mm, and 12 ± 0.4 mm	Omar et al. (2017)
Portieria hornemannii	AgNPs	Beauveria bassiana and Metarhizium anisopliae	100 µL	22.6 mm & 21 mm	Ramamoorthy et al. (2019)

(continued)

1.9 Applications of Marine Flora-Derived Nanoparticles (MFNPs)

Table 1.2 (continued)

Source	NPs	Inhibitory activity against	Dose	Inhibition zone/level	Reference
Pterocladia capillacea	Fe$_3$O$_4$NPs	*Aspergillus fumigatus* and *Fusarium moniliforme*	30 mg/mL	5 mm and 3 mm	Salem et al. (2019)
Rhodymenia palmata	AgNPs	*Aspergillus flavus, Aspergillus fumigatus*	2 μg	3 ± 0.76 mm and 5 ± 0.06 mm	Murugammal and Flora (2017)
Seagrass		**Bacteria**			
Cymodocea serrulata	AgNPs	*Vibrio parahaemolyticus* MTCC 451	100 μL-LD$_{50}$ dose	15.2% of shrimp mortality	RathnaKumari et al. (2018)
Halophila stipulacea	AgNPs	*Oscillatoria simplicissima*	4 μL	95 ± 0.05%	El-Kassas and Ghobrial (2017)
Syringodium isoetifolium	AgNPs	*S. mutans*	0.24 nm	14.3 ± 0.12 mm	Ahila et al. (2016)
		Fungi			
Cymodocea serrulata	AgNPs	*Pyriporia oryzea, Helminthisporium oryzea, Alternaria* sp	100 μL	14 ± 1.5 mm, 12 ± 1.4 mm, and 10 ± 1.2 mm	Kailasam et al. (2023)
Mangroves		**Bacteria**			
Avicennia alba	AgNPs	*Agrobacterium tumefaciens, Escherichia coli, Staphylococcus aureus,* and *Streptococcus mutans*	10 ppm	12.2 ± 0.7, 8.3 ± 0.6, 11 ± 0.6, 10.12 ± 0.25	Bakshi et al. (2015)

(continued)

Table 1.2 (continued)

Source	NPs	Inhibitory activity against	Dose	Inhibition zone/level	Reference
Avicennia alba	AgNPs	*Micrococcus luteus* MTCC 106, *Arthrobacter protophormiae* MTCC 2682, *Rhodococcus rhodochrous* MTCC 265, *Enterococcus faecalis* MTCC 439, *Streptococcus mutans* MTCC 497, *Bacillus subtilis* MTCC 441, *Enterobacter aerogenes* MTCC 10208, *Alcaligens faecalis* MTCC 126, *Proteus vulgaris* MTCC 426, *Proteus mirabilis* MTCC 425, *Pseudomonas aeruginosa* MTCC 1688, *and Salmonella enteric* MTCC 3858	100 μL	4–17 mm	Nagababu and Rao (2016)
Avicennia marina	AgNPs	*E. coli, Pseudomonas aeruginosa, Staphylococcus aureus, Klebsiella* sp., and *Bacillus subtilus*	6.25 μg mL^{-1} MIC, 12.5 μg mL^{-1} MIC, 25 μg mL^{-1} MIC, 6.25 μg mL^{-1} MIC, and 25 μg mL^{-1} MIC	18.40 ± 0.97 mm, 17.64 ± 0.91 mm, 10.87 ± 1.33 mm, 14.65 ± 1.09 mm, and 13.93 ± 0.84 mm	Gnanadesigan et al. (2012)
Avicennia marina	AgNPs	*E. coli, K. pneumoniae* and *P. aeruginosa*	6.25 μg/mL, 12.5 μg/mL and 12.5 μg/mL	–	Naidu et al. (2019)
Avicennia marina	TiO$_2$NPs	*Micrococcus* sp., *Staphylococcus aureus*, and *Escherichia coli*	20 μL	13/66 ± 1/55 mm, 12 ± 2/66 mm, and 12 ± 0/66	Lefteh et al. (2020)
Ceriops decandra	AgNPs	*L. monocytogenes, E. coli* and *S. typhimurium*	50 μg/mL	47.93 ± 5.98, 10.11 ± 0.18 and 28.17 ± 5.97%	Maity et al. (2019)
Ceriops decandra	AgNPs	TGBS09	150 μg/disk	18 mm	Sankar and Abideen (2019)

(continued)

1.9 Applications of Marine Flora-Derived Nanoparticles (MFNPs)

Table 1.2 (continued)

Source	NPs	Inhibitory activity against	Dose	Inhibition zone/level	Reference
Ceriops tagal	AgNPs	Staphylococcus aureus, Pseudomonas aeuroginosa, Escherichia coli, Bacillus cereus, and Proteus mirabilis	20 μL	14 mm, 21 mm, 11 mm, 13 mm, and 11 mm	Dhas et al. (2013)
Excoecaria agallocha	AgNPs	Salmonella typhi, Pseudomonas aeruginosa, and Staphylococcus aureus	100 μL	—	Bhuvaneswari et al. (2017)
Heritiera fomes	AgNPs	S. aureus, V. cholera, S. epidermidis, B. subtilis, and	100 μg NPs/disk	12 mm, 9 mm, 10 mm, 15 mm, and 14 mm	Thatoi et al. (2016)
Lumnitzera racemosa	AgNPs	Escherichia coli, Staphylococcus aureus, Bacillus subtilis, and Klebsiella pneumoniae	200 mg/mL	13 ± 0.6 mm, 16 ± 0.3 mm, 07 ± 0.9 mm, and 10 ± 0.8 mm	Arshan et al. (2020)
Rhizophora apiculata	AgNPs	Bacillus subtilis, Pseudomonas aeruginosa, Escherichia coli, Klebsiella pneumoniae, Proteus vulgaris, Salmonella typhi, and Staphylococcus aureus	50 μL	11 mm, 12 mm, 14 mm, 14 mm, 14 mm, 14 mm, and 14 mm	Antony et al. (2011)
Rhizophora mucronata	AgNPs	Proteus spp., Pseudomonas fluorescens and Flavobacterium spp.	75 μg/μL	14 mm, 16 mm, and 14 mm	Umashankari et al. (2012)
Rhizophora mucronata	AgNPs	Bacillus cereus, Staphylococcus aureus, Vibrio harveyi, & Pseudomonas aeruginosa	20 μL of 10 mg/mL AgNPs	18.33 ± 0.57, 17.66 ± 1.52, 13.66 ± 0.57, & 15.33 ± 0.57	Abdi et al. (2019a)
Rhizophora mucronata	AgNPs	Escherichia coli, Staphylococcus aureus, and Pseudomonas aeruginosa	50, 25 and 25 μg/mL	—	Premanathan et al. (2014)

(continued)

Table 1.2 (continued)

Source	NPs	Inhibitory activity against	Dose	Inhibition zone/level	Reference
Rhizophora mucronata	AgNPs	TGBS09	150 µg/disk	18 mm	Sankar and Abideen (2019)
Rhizophora mucronata	AgNPs	Bacillus cereus, Staphylococcus aureus, Vibrio harveyi, and Pseudomonas aeruginosa	20 µL (10 mg/1 mL)	Leaf: 12.33 ± 0.57 mm, 14 ± 1.73 mm, 10.33 ± 1.52 mm, and 7.66 ± 1.52; Root: 15.66 ± 3.05 mm, 17 ± 1.73 mm, 14 ± 2.00 mm, and 12.66 ± 2.08 mm; Stem: 18.33 ± 0.57 mm, 17.66 ± 1.52 mm, 13.66 ± 0.57 mm, and 15.33 ± 0.57 mm	Abdi et al. (2019b)
Rhizophora stylosa	AgNPs	E scherichia coli and Staphyloccocus aureus	100 mg/mL	7.2 mm and 4.3 mm	Willian et al. (2022)
Rhizophora stylosa	AgNPs	E scherichia coli and Staphyloccocus aureus	10 mM	> 8 mm	Willian et al. (2020)
Sonneratia alba	AgNPs	Bacillus subtilis, Klebsiella pneumoniae, and Salmonella typhi	150 mg/L	18.45 ± 0.95 mm, 18.85 ± 0.74 mm, and 21.22 ± 1.35 mm	Murugan et al. (2017)
Sonneratia apetala	AgNPs	Agrobacterium tumefaciens, Escherichia coli, Staphylococcus aureus, and Streptococcus mutans	10 ppm	11 ± 0.8, 11 ± 0.3, 9.2 ± 0.7, and 8.63 ± 0.25	Bakshi et al. (2015)
Sonneratia apetala	AgNPs	Proteus mirabilis	100 µL of 100 mg/mL	27.3 mm	Nagababu and Rao (2017)
Sonneratia apetala	AgNPs	S. aureus, S. flexneri, V. cholera, S. epidermidis, B. subtilis, and E. coli	100 µg NPs/disk	16 mm, 13 mm, 10 mm, 12 mm, 14 mm, and 10 mm	Thatoi et al. (2016)
Sonneratia apetala	ZnONPs	S. flexneri	100 µg NPs/disk	11 mm	Thatoi et al. (2016)

(continued)

1.9 Applications of Marine Flora-Derived Nanoparticles (MFNPs)

Table 1.2 (continued)

Source	NPs	Inhibitory activity against	Dose	Inhibition zone/level	Reference
Sonneratia caseolaris	AgNPs	Agrobacterium tumefaciens, Escherichia coli, Staphylococcus aureus, and Streptococcus mutans	10 ppm	7.5 ± 0.4, 9.6 ± 0.8, 7.9 ± 0.5, and Nil	Bakshi et al. (2015)
Xylocarpus granatum	AgNPs	Escherichia coli ATCC 25,922, Salmonella typhimurium, Salmonella typhi MTCC 734	1500 µg/mL	1.5 cm, 1.3 cm, and 1.1 cm	Maity and Mondal (2017)
		Fungi			
Avicennia alba	AgNPs	Aspergillus flavus and Trichophyton rubrum	1–10 ppm	6.5–10.5 mm and 6.2–7 mm	Bakshi et al. (2015)
Ceriops tagal	AgNPs	Aspergillus niger, Candida albicans, and Candida tropicalis	20 µL	8 mm, 7 mm, and 8 mm	Dhas et al. (2013)
Rhizophora apiculata	AgNPs	Candida albicans	20 µL	24–26 mm	Manoj Singh et al. (2013a, b)
Sonneratia apetala	AgNPs	Aspergillus flavus and Trichophyton rubrum	1–10 ppm	6.9–10.5 mm and 7.8–8.5 mm	Bakshi et al. (2015)
Sonneratia caseolaris	AgNPs	Aspergillus flavus and Trichophyton rubrum	1–10 ppm	6.5–8.9 mm and 8.5 mm	Bakshi et al. (2015)
Sea-blites and other plants associated with marine plants		**Bacteria**			

(continued)

Table 1.2 (continued)

Source	NPs	Inhibitory activity against	Dose	Inhibition zone/level	Reference
Clerodendrum inerme	AgNPs	Pseudomanas aeruginosa, Klebsiellapneumoniae, Staphylococcus aureus, Listeria monocytogenes, and Micrococcus luteus	50 µL	17 ± 1.8 mm, 20 ± 2.2 mm, 15 ± 1.7 mm, 21 ± 2.3 mm, and 20 ± 2.2 mm	Kathiresan et al. (2012)
Hibiscus tiliaceus	AgNPs	X. campestris, R. solanacearum	200 µg/mL	26.6 ± 0.25 mm and 24.2 ± 0.69 mm	Usha Rani et al. (2016)
Ipomoea pes-caprae	AgNPs	S. aureus, P. aeruginosa and E. coli	100 µL	25 mm, 23 mm, and 22 mm	Subha et al. (2015)
Ipomoea pes-caprae	AgNPs	E. coli, P. aeruginosa. K. pneumonia, Enterobacter sp., and S. aureus	0.5% AgNP solution	8.25 ± 0.65 mm, 6.15 ± 0.25 mm, 6.30 ± 0.72 mm, 7.20 ± 0.46 mm, and 6.50 ± 0.280 mm	Satyavani et al. (2013)
Ipomoea pes-caprae	AgNPs	P. aeruginosa, E. coli and Bacillus	100 µg	13 ± 2 mm, 19 ± 2 mm, 14 ± 1 mm	Veeramani et al. (2018)
Sesuvium portulacastrum	AgNPs	Pseudomanas aeruginosa, Klebsiellapneumoniae, Staphylococcus aureus, Listeria monocytogenes, and Micrococcus luteu	50 µL	20 ± 2.4 mm, 20 ± 2.1 mm, 18 ± 1.9 mm, and 18 ± 2.0 mm	Kathiresan et al. (2012)
Sesuvium portulacastrum	AgNPs	Staphylococcus aureus and Micrococcus luteus	50 µL/disk	23 mm and 8 mm	Nabikhan et al. (2010)
		Fungi			
Clerodendrum inerme	AgNPs	Alternaria alternata, Candida albicans, Penicillium italicum, and Fusarium equisetii	50 µL	15 ± 1.4 mm, 12 ± 1.5 mm, 10 ± 1.3 mm, and 07 ± 1.1 mm	Kathiresan et al. (2012)

(continued)

1.9 Applications of Marine Flora-Derived Nanoparticles (MFNPs)

Table 1.2 (continued)

Source	NPs	Inhibitory activity against	Dose	Inhibition zone/level	Reference
Sesuvium portulacastrum	AgNPs	Alternaria alternata, Candida albicans, Penicillium italicum, and Fusarium equisetii	50 μL	16 ± 1.8 mm, 13 ± 1.5 mm, 17 ± 1.9 mm, and 17 ± 1.9 mm	Kathiresan et al. (2012)
Sesuvium portulacastrum	AgNPs	Penicillium and Candida albicans	50 μL/disk	18 mm and 12 mm	Nabikhan et al. (2010)

18.06 ± 0.25 mm inhibition zones, respectively (Anjali et al. 2022). AgNPs fabricated from *Codium tomentosum* at 150 mg/mL dose resulted inhibition of *Bacillus subtilis, Klebsiella pneumoniae*, and *Salmonella typhi* with zone of inhibitions of 18.36 ± 1.02 mm, 17.68 ± 0.28 mm, and 20.24 ± 0.84 mm, correspondingly (Murugan et al. 2016b). *Enteromorpha flexuosa* derived AgNPs with 12.5, 6.25, 50, 25, 6.25, 50, and 50 μg mL^{-1} concentrations inhibited *B. subtilis, B. pumulis, E. faecalis, S. aureus, S. epidermidis, E. coli,* and *K. pneumoniae* with 18 ± 0.8, 19 ± 1.2, 12 ± 0.9, 14 ± 0.7, 20 ± 1.5, 13 ± 0.9, and 10 ± 0.4 mm inhibition zones, respectively (Yousefzadi et al. 2014). *Ulva lactuca* originated AgNPs at 1 mg concentration resulted inhibition of *Bacillus subtilus, Escherichia coli, Klebsiella* sp., *Pseudomonas aeuroginosa, Staphylococcus aureus, Streptococcus faecalis,* and *Neisseria gonorrhoeae,* with 30 mm, 26 mm, 17 mm, 27 mm, 25 mm, 13 mm and 11 mm zone of inhibition (Amin 2020). AgNPs green synthesized from *Ulva reticulata* at 50 μL dose displayed 26 mm and 25 mm zone of inhibitions against *Bacillus* sp.and *Staphylococcus aureus* (Bhimba and Devi 2014). FeNPs synthesized from *Ulva lactuca* demonstrated potential antibacterial activity against *E. coli, S. aureus, S. typhimurium,* and *P. vulgaris* at 20 μL concentration (Bensy et al. 2022). SeNPs synthesized from *U. lactuca* at 100 μL/mL concentration inhibited *Lactobacillus, S. aureus,* and *S. mutans* with > 30 mm, > 15 mm, and > 25 mm zone of inhibitions (Vikneshan et al. 2020).

Considerably, Ochrophyta and Rhodophyta species have been widely used to derive a wide spectrum of nanoparticles (Fig. 1.7) due to their availability as well as well-known biological properties. *Cystoseira crinita* derived ZnONPs at 100 μL (10 mg mL^{-1}) concentration showed 26.0 ± 0.57 mm, 27.3 ± 0.33 mm, 24.3 ± 0.33 mm, and 23.6 ± 0.88 mm inhibition zones against *Bacillus cereus, Staphylococcus aureus, Escherichia coli,* and *Salmonella typhi* (Elrefaey et al. 2022). AuNPs synthesized from *Ecklonia cava* demonstrated 31.8 ± 0.32 mm, 19.7 ± 0.21 mm, 21.3 ± 0.28 mm, and 16.6 ± 0.30 mm inhibition zones against *E. coli, B. subtilis, P. aeruginosa* and *S. aureus* at 20 μL volume concentration (Venkatesan et al. 2014). AgNPs synthesized from *Saccharina japonica* at 100 μL concentration resulted 20 ±

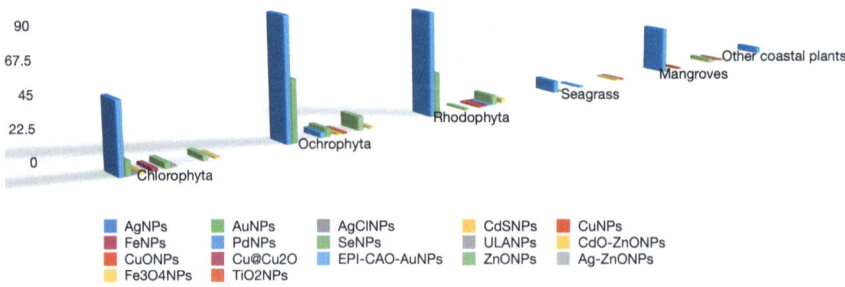

Fig. 1.7 Number of reports on nanoparticles synthesized from each marine floral group

1.9 Applications of Marine Flora-Derived Nanoparticles (MFNPs)

0.70 mm, 16 ± 0.10 mm, 16 ± 0.25 mm, 14 ± 0.25 inhibition zones against *E. coli*, *S. typhimurium*, *L. monocytogenes*, *B. cereus*, and *S. aureus* (Sivagnanam et al. 2017). The genus *Sargassum* has been widely exploited than other marine flora to synthesize nanoparticles and demonstrated their applications in antimicrobial research (Table 1.2). AgNPs derived from *Sargassum ilicifolium* (Kumar et al. 2012a), *Sargassum myriocystum* (Balaraman et al. 2020), *Sargassum polycystum* (Thangaraju et al. 2012) (Thiurunavukkarau et al. 2022), *Sargassum polyphyllum* (Arunkumar et al. 2014), and *Sargassum wightii* (Selvaraj et al. 2020) (Thiurunavukkarau et al. 2022), showed potential antibacterial activity against numerous pathogenic bacteria. *Sargassum angustifolium* extract coated SENPs demonstrated anti-vibriosis activity by inhibiting *Penaeus vannamei* pathogen (aquaculture industry interested pathogen) -*Vibrio harveyi* at 200 μg/mL MIC (Mansouri-Tehrani et al. 2021). *Sargassum wightii* originated ZrO_2NPs at 15 μg concentration resulted 21 mm, 19 mm, and 19 mm inhibition zones against *B. subtilis, E. coli,* and *S. typhi* (Kumaresan et al. 2018). *Padina boryana* derived PdNPs at 100 μL of 1 mg/mL concentration inhibited *S. aureus* (18.3 ± 1.24 mm), *Escherichia fergusonii* (20.0 ± 0.81 mm), *Acinetobacter pittii* (23.0 ± 0.8 mm), *P. aeruginosa* (21.3 ± 0.47 mm), *Aeromonas enteropelogenes* (19.3 ± 0.5 mm), and *Proteus mirabilis* (23.0 ± 1.6 mm) (Sonbol et al. 2021).

Silver nanocomposites synthesized from the aqueous extracts of red seaweeds have demonstrated bactericidal activity against various plant pathogens as promising nanopesticides (Roseline et al. 2019). *Acanthophora spicifera* derived AgNPs at 50 μL of 100 μg/mL concnetration inhibited *M. luteus, S. aureus, P. fluorescens, E. coli, Serratia marcescens, K. pneumoniae,* and *B. subtilis,* with zone of inhibitions 36 mm, 36 mm, 24 mm, 18 mm, 16 mm, 17 mm, 17 mm and 14 mm, correspondingly (Thiurunavukkarau et al. 2022). AgNPs synthesized from *Champia parvula* demonstrated 23 mm, 21 mm, and 20 mm inhibition zones at 100 μL/mL concentration against *S. mutans, S. aureus, and Enterococcus faecalis* (Viswanathan et al. 2023). *Gracilaria crassa* originated AgNPs at 40 μg/mL showed ~ 42 mm, ~ 38 mm, ~ 22 mm, and ~ 23 mm inhibition zones against *E. coli, P. mirabilis, B. subtilis* and *P. aeruginosa* (Lavakumar et al. 2015). AgNPs derived from other red algal species such as *Halymenia porphyroides* (Kiran and Murugesan, 2014b), *Hypnea musciformis* (Devi and Bhimba 2014; Vadlapudi and Amanchy 2017), *Hypnea valentiae* (Viswanathan et al. 2022), *Laurencia papillosa* (Omar et al. 2017), *Pterocladiella capillacea* (Kassas and Attia 2014), and *Spyridia fusiformis* (Murugesan et al. 2017), have also demonstrated potential inhibitory action against various other pathogens. AuNPs synthesized from *Halymenia pseudofloresii* extracts displayed inhibitory action against *S. aureus* (24 mm), *Lactobacillus* (23 mm), and *P. aeruginosa* (22 mm) at 100 μg/mL dose (Palaniyandi et al. 2023). Co_3O_4NPs derived from a marine red alga showed 21.1 ± 7.1 mm, 23.6 ± 6.9 mm, and 20.8 ± 5.8 mm inhibition zones against *S. aureus, B. subtilis, E. coli,* and *P. aeruginosa* at 25.0 ± 7.3 μg/mL MIC concentration (Ajarem et al. 2022).

AgNPs green synthesized from seagrass *Halophila stipulacea* displayed 95 ± 0.05% inhibitory activity against *Oscillatoria simplicissima* at 4 μL concentration (El-Kassas and Ghobrial 2017). AgNPs fabricated from mangrove *Avicennia marina* inhibited *E. coli* (18.40 ± 0.97 mm), *P. aeruginosa* (17.64 ± 0.91 mm),

S. aureus (10.87 ± 1.33 mm), *Klebsiella* sp. (14.65 ± 1.09 mm), and *B. subtilus* (13.93 ± 0.84 mm) at specific concentrations of 6.25 μg mL^{-1} MIC, 12.5 μg mL^{-1} MIC, 25 μg mL^{-1} MIC, 6.25 μg mL^{-1} MIC, and 25 μg mL^{-1} MIC, respectively (Gnanadesigan et al. 2012). *Rhizophora mucronata* originated AgNPs at 20 μL of 10 mg/mL inhibited *Bacillus cereus* (18.33 ± 0.57 mm), *S. aureus* (17.66 ± 1.52), *V. harveyi* (13.66 ± 0.57), and *P. aeruginosa* (15.33 ± 0.57) (Abdi et al. 2019a). AgNPs derived from mangrove, *R. mucronata,* inhibited different fish pathogens (Umashankari et al. 2012). AgNPs obtained from *Sonneratia alba* extracts demonstrated inhibitory action at 150 mg/L against *B. subtilis* (18.45 ± 0.95 mm), *K. pneumoniae* (18.85 ± 0.74 mm), and *S. typhi* (21.22 ± 1.35 mm) (Murugan et al. 2017).

Silver nanoparticles derived from coastal plants like *Clerodendrum inerme* showed 17 ± 1.8 mm, 20 ± 2.2 mm, 15 ± 1.7 mm, 21 ± 2.3 mm, and 20 ± 2.2 mm at 50 μL dose against *P. aeruginosa, K. pneumoniae, S. aureus, Listeria monocytogenes*, and *M. luteus*, respectively (Kathiresan et al. 2012). *Hibiscus tiliaceus* derived AgNPs at 200 μg/mL concentration inhibited *X.* campestris (26.6 ± 0.25 mm) and *R. solanacearum* (24.2 ± 0.69 mm) (Usha Rani et al. 2016). AgNPs synthesized from *Ipomoea pes-caprae* at 100 μL dose displayed 25 mm, 23 mm, and 22 mm inhibition zones against *S. aureus, P. aeruginosa* and *E. coli* (Subha et al. 2015). AgNPs (50 μL) fabricated from *Sesuvium portulacastrum* extracts showed 20 ± 2.4 mm, 20 ± 2.1 mm, 18 ± 1.9 mm, and 18 ± 2.0 mm inhibition zones against *P. aeruginosa, K. pneumoniae, S. aureus, L. monocytogenes*, and *M. luteus* (Kathiresan et al. 2012).

1.9.3 Antituberculosis Activity

S. polycystum derived AgNPs were reported to show anti-mycobacterial activity against *Mycobacterium tuberculosis* strains MTBH37Rv (99.38% inhibition), MTB all drug sensitive (94.79%) and MTD MTB at 100 μg/mL concentration (82.44%) (Thiurunavukkarau et al. 2022). Nanoparticle derived from other marine plants have not been tested to investigate their antituberculosis activity, indicating research lacuna in this application.

1.9.4 Antifungal Properties

A limited subset of marine plant-derived nanoparticles exhibited pronounced antifungal efficacy against a diverse range of fungal pathogens. AgNPs fabricated from green macroalgae *Ulva lactuca* inhibited *Fusarium oxysporum* (54.33 ± 1.53 mm), *Botrytis cinerea* (71.00 ± 2.00 mm), and *Colletotrichum gloeosporioides* (70.33 ± 2.08 mm) at 30 μL concentration (Koçer and Özçimen 2022). ZnONPs fabricated from *Ulva lactuca* at 100 μg/mL dose inhibited *B. licheniformis* (26.3 ± 1.6 mm), *B. pumilus* (21.2 ± 0.9 mm), *E. coli* (24.0 ± 1.0 mm), and *P. vulgaris* (20.3 ± 0.7 mm)

1.9 Applications of Marine Flora-Derived Nanoparticles (MFNPs)

(Ishwarya et al. 2018a). *Ulva reticulata* originated AgNPs at 50 μL concentration displayed 36 mm, 30 mm and 30 mm inhibition zones against *Candida albicans, Candida parapsilosis* and *Aspergillus niger,* respectively (Bhimba and Devi 2014). *Ulva rigida* AgNPs at 32 μg mL^{-1} MLC displayed 30 mm and 40 mm inhibition against *T. cataneum* and *T. mantigrophytes* (Algotiml et al. 2022b). AuNPs synthesized from *Ulva rigida* showed 30 ± 1 mm and 25 mm inhibition zones at 64 μg mL^{-1} MLC against *T. cataneum* and *T. mantigrophytes,* respectively (Algotiml et al. 2022a).

ZnONPs (100 μL 10 mg mL^{-1}) fabricated from brown seaweed *Cystoseira crinita* displayed 22.3 ± 0.33 mm and 31.0 ± 0.57 mm inhibition against *C. albicans and A. niger* (Elrefaey et al. 2022*). Cystoseira myrica* originated AgNPs at 32 μg mL^{-1} MLC dose demonstrated 30 mm and 30 mm inhibition zones against *T. cataneum and T. mantigrophytes* (Algotiml et al. 2022b). *Ecklonia cava* AuNPs (20 μL) showed 24.6 ± 0.23 mm, 19.3 ± 0.26 mm, 21.5 ± 0.25 mm, and 23.3 ± 0.25 mm inhibitory zones against *Aspergillus niger, A. brasiliensis, A. fumigates,* and *C. albicans,* respectively (Venkatesan et al. 2014). AgNPs obtained from *Halopteris scoparia* extracts inhibited *Fusarium oxysporum, Botrytis cinerea,* and *Colletotrichum gloeosporioides* at 30 μL concentration with zone of inhibitions 49.00 ± 1.73 mm, 70.33 ± 2.08 mm, and 68.67 ± 2.89 mm, respectively (Koçer and Özçimen, 2022). *Padina tetrastromatica* derived AgNPs at 150 μL dose showed 20.03 ± 0.033 mm, and 18.13 ± 0.089 mm, 12.20 ± 0.152 mm, 12.20 ± 0.100 mm, and 10.17 ± 0.167 mm inhibition zones against *Fusarium* sp., *Aspergillus niger, Candida albicans, Aspergillus fumigatus,* and *Aspergillus* sp. (Rajeshkumar et al. 2017b). CuONPs (50 μg/mL) synthesized from *Sargassum polycystum* revealed 20 ± 0.5 mm and 12 ± 0.5 mm inhibition against *A. niger* and *A. oryzae* (Ramaswamy et al. 2016). *Spatoglossum asperum* derived AgNPs at 100 μg/mL concentration inhibited *Candida albicans, Candida tropicalis, Trichophyton mentagrophytes, and Aspergillus flavus* with 20.67 ± 0.88 mm, 17.67 ± 0.33 mm, 17.33 ± 0.88 mm, and 12.67 ± 0.33 mm inhibition zones (Subbiah et al. 2019).

Red algae *Acanthophora spicifera* derived AgNPs at 50 μL of 100 μg/mL concentration demonstrated inhibition (20 mm) of *Candida albicans* (Thiurunavukkarau et al. 2022). AgNPs fabricated from *Champia parvula* at 100 μL/mL dose caused 20 mm inhibition zone against *C. albicans* (Viswanathan et al. 2023). *Hypnea musciformis* derived AgNPs at 80 μL volume inhibited *A. niger* (27 mm), *C. albicans* (26 mm) and *C. parasilopsis* (21 mm) (Devi and Bhimba, 2014). AgNPs (100 μL) synthesized from *Portieria hornemannii* showed 22.6 mm and 21 mm inhibition zones against *Beauveria bassiana and Metarhizium anisopliae* (Ramamoorthy et al. 2019).

Seagrass *Cymodocea serrulata* synthesized AgNPs at 100 μL displayed 14 ± 1.5 mm, 12 ± 1.4 mm, and 10 ± 1.2 mm inhibition zones against *Pyriporia oryzea, Helminthisporium oryzea,* and *Alternaria* sp. (Kailasam et al. 2023). Mangrove *Rhizophora apiculata* originated AgNPs (20 μL) showed 24–26 mm inhibition of *C. albicans* (Manoj Singh et al. 2013a, b). The sea purslane *Sesuvium portulacastrum* derived AgNPs at 50 μL dose displayed 16 ± 1.8 mm, 13 ± 1.5 mm, 17 ± 1.9 mm, and 17 ± 1.9 mm inhibition zones against *Alternaria alternata, C. albicans, Penicillium italicum,* and *Fusarium equisetii* (Kathiresan et al. 2012) and 18 mm and 12 mm against *Penicillium* and *C. albicans* (Nabikhan et al. 2010).

1.9.5 Anticancer and Cytotoxic Properties

Nanoparticles derived from microbes and terrestrial plants are reported to demonstrate promising anticancer activity for further drug development endeavors (Dan Zhang et al. 2020a, b). While, marine plants being potential drug resources from sea, offer much effective nanodrugs of green synthesized nanoparticles with potential anticancer activity. Nanoparticles originated from marine plant extracts have demonstrated promising cytotoxic activity against a variety of cancer cell lines. AuNPs have been thought to involve in DNA damage, genotoxicity inflammation, mitochondrial damage, protein denaturation, membrane damage, apoptosis, and inflammation cytotoxicity (Baskar et al. 2023). A similar mechanism is proposed for Fe_3O_4NPs' anticancer activity (Yew et al. 2020). The adhesion of MgONPs with electron transportation chain and cell wall membrane facilitates arrest of permeability (Fouda et al. 2022b). Subsequently they enter inside the cells, enhances ROS levels, inhibits electro transportation chain, oxidative damage, and react with proteins, enzymes, plasmids, and DNA to cause disruption of cell function thereby inhibiting cells (Fouda et al. 2022b) (Fig. 1.6).

Among numerous known marine plants, seaweeds *Ulva, Cystoseira, Padina, Sargassum, Gracilaria,* and seagrass *Cymodocea* (Table 1.3; Figs. 1.8, 1.9, 1.10, 1.11, 1.12 and 1.13), are used often for nanoparticles as anticancer nanodrugs (Table 1.3). Most of the investigated nanoparticles originated from marine plants have demonstrated considerable anticancer activity against a wide variety of cancer cell lines. Silver nanoparticles appear to have more demand over other nanoparticles due to cytotoxic activity (Table 1.3). A lowest dosage but highly efficient to demonstrate anticancer activity against various cell lines was reported from very few nanoparticles derived marine flora (Table 1.1 and 1.3). AgNPs derived from green seaweed *Ulva lactuca* showed 50% inhibition of Hep-2 cell lines at 12.5 µg/mL IC_{50} concentration (Devi and Bhimba 2012). SeNPs synthesized from *Ulva fasicata* showed 50% anticancer activity against MDA-MB-231 cell lines at 11 µg/mL IC_{50} concentration (Shahzamani et al. 2022) (Figs. 1.9, 1.10, 1.11, and 1.12).

Brown seaweeds, due to their natural bioactive nature, have been widely used to synthesize a wide variety of nanoparticles, particularly for AuNPs, for their anticancer properties compared to any other marine plants (Table 1.3; Fig. 1.7). Among brown seaweeds, *Cystoseira myrica* fabricated AgNPs demonstrated 50% inhibition of HFb-4 and MCF-7 cell lines at 13 µg/mL IC_{50} (Algotiml et al. 2022b). *Fucus vesiculosus* derived DOX–AcFu NPs displayed 50% inhibitory action against HCT-116 and HCT-8 cell lines at ~1 µg/mL IC_{50} dose (Lee et al. 2013). AuNPs fabricated from *Padina tetrastromatica* inhibited (47%) HepG2 and A549 cell lines at 100 µg concentration (Rajeshkumar et al. 2017a, b). Similarly, AuNPs synthesized from

1.9 Applications of Marine Flora-Derived Nanoparticles (MFNPs)

Table 1.3 Anticancer and cytotoxic activity of marine plants derived nanoparticles against different cell lines

Nanoparticles derived from marine plants	NPs	Inhibitory Dose	Cell lines	Inhibition activity	Reference
Green seaweed					
Caulerpa scalpelliformis	AgNPs	40 μg/mL	MCF-7 cells	50%	Manikandan et al. (2019)
Caulerpa taxifolia	AgNPs	40 μg/mL	A549 lung cancer cell	~52%	Danjie Zhang et al. (2020a, b)
Enteromorpha compressa	AgNPs	95.35 μg mL^{-1} IC$_{50}$	Ehrlich Ascites Carcinoma cells	50%	Vijayan Sri Ramkumar et al. (2017a, b)
Ulva fasicata	Ag/AgClNPs	7.38 μg/mL	A549 cell line	50%	Lashgarian et al. (2021)
Ulva fasicata	SeNPs	11 μg/mL IC$_{50}$	MDA-MB-231	50%	Shahzamani et al. (2022)
Ulva fasicata	AgNPs	98 μg/mL	Ehrlich Ascites Carcinoma (EAC)	94%	Khalifa et al. (2016)
Ulva lactuca	AgNPs	12.5 μg/mL, 37 μg/mL, 49 μg/mL, and 95 μg/mL IC$_{50}$	Hep-2, MF7, HT29 and Vero cell lines	50%	Devi and Bhimba (2012)
Ulva lactuca	AuNPs	23 μM IC$_{50}$	HT-29	50%	González-Ballesteros et al. (2019a, b)
Ulva lactuca	AgNPs	13 μM IC$_{50}$	HT-29	50%	González-Ballesteros et al. (2019a, b)
Ulva lactuca	ULANPs	10 μM IC$_{50}$	HepG2 and MCF7 cells	48.31 + 3.4% and 50.39 + 2.4%	Al-Malki (2020)
Ulva lactuca	AgNPs	200 μM	HCT-116 cells	90%	Acharya et al. (2022)
Ulva lactuca	FeNPs	157.366 mg/mL LC$_{50}$	HeLa and DLD-1	50%	Bensy et al. (2022)
Ulva rigida	AgNPs	13 μg mL^{-1} IC$_{50}$	HFb-4 and MCF-7 cell lines	50%	Algotiml et al. (2022b)

(continued)

Table 1.3 (continued)

Nanoparticles derived from marine plants	NPs	Inhibitory Dose	Cell lines	Inhibition activity	Reference
Ulva rigida	AuNPs	30 μg mL^{-1} IC$_{50}$	MCF-7 cell line	86.83%	Algotiml et al. (2022a)
Brown seaweed					
Chondrus crispus	AuNPs	50 mg/mL	A549, THP-1, and HL-60 cell lines		González-Ballesteros et al. (2022)
Codium tomentosum	AuNPs	40 μM	HEPG-3 and BxPC-3 cells	60%	González-Ballesteros et al. (2023)
Colpomenia sinuosa	AgNPs	200 μg/mL	DLA and EAC cell lines	61.57 ± 3.45% and 81.96 ± 8.52%	Vishnu and Murugesan, (2014)
Cystoseira baccata	AuNPs	79.03 μM and 49.61 μM IC$_{50}$	Caco-2 and HT- 29 cell lines	50%	González-Ballesteros et al. (2017)
Cystoseira crinita	MgONPs	113.4 μg mL^{-1} and 141.2 μg mL^{-1} IC$_{50}$	Caco-2 and Vero cancer cell lines	50%	Fouda et al. (2022a)
Cystoseira myrica	AuNPs	51 μg mL^{-1} IC$_{50}$	MCF-7 cell line	89.92%	Algotiml et al. (2022a)
Cystoseira myrica	AgNPs	13 μg mL^{-1} IC$_{50}$	HFb-4 and MCF-7 cell lines	50%	Algotiml et al. (2022b)
Cystoseira myrica	AgNPs	100 μg/mL	MCF-7 breast carcinoma cells and HepG2 human hepatocellular carcinoma cells	80.1% and 74.19%	Mohamed et al. (2022)
Cystoseira myrica	AgNPs	98 μg/mL	Ehrlich Ascites Carcinoma (EAC)	83%	Khalifa et al. (2016)
Dictyota bartayresiana	AgNPs	296.14 μL/l	DLA cells	50%	Antonysamy et al. (2015)
Ecklonia cava	AuNPs	20 μg/mL	HaCaT cells	~90%	Venkatesan et al. (2014)

(continued)

1.9 Applications of Marine Flora-Derived Nanoparticles (MFNPs)

Table 1.3 (continued)

Nanoparticles derived from marine plants	NPs	Inhibitory Dose	Cell lines	Inhibition activity	Reference
Ecklonia cava	AgNPs	59 µg/mL IC_{50}	HeLa cells	50%	Venkatesan et al. (2016)
Fucus evanescens	AgNPs	50 µg/mL IC_{50}	Rat C6 glioma cells	+	Yugay et al. (2020)
Fucus vesiculosus	DOX–AcFu NPs	~1 µg/mL IC_{50}	HCT-116 and HCT-8	50%	Lee et al. (2013)
Fucus vesiculosus	Fu/CHNPs	8×10^9 NPs/mL	Human endothelial and breast cancer cells	-	Oliveira et al. (2018)
Fucus vesiculosus	PEI-FCD-DOXNPs	0.1770 ± 0.092 µM IC_{50}	MDA-MB-231 cells	50%	Pawar et al. (2019)
Laminaria digitata	ZnONPs	20.69 µg/mL IC_{50} and 16.21 µg/mL IC_{50}	Human dermal fibroblast (HDF) and human colon cancer (HT-29) cells	50%	Vijayakumar et al. (2022)
Padina boryana	PdNPs	125 µg/mL	MCF-7	53%	Sonbol et al. (2021)
Padina gymnospora	AuNPs	82.91 nM and 144.16 nM IC_{50}	HepG2 and A549 cell lines	50%	Singh et al. (2014)
Padina gymnospora	AuNPs	76.40 nM IC_{50}	HepG2 and A549 cell lines	50%	Singh et al. (2015)
Padina tetrastromatica	AuNPs	100 µg	HepG2 and A549 cell lines	47%	Rajeshkumar et al. (2017a, b)
Padina tetrastromatica	AgNPs	86.7 µg mL^{-1} IC_{50}	MCF-7 cell lines	50%	Selvi et al. (2016)
Polycladia myrica	SeNPs	123.51 ± 4.07 g/mL and 220.53 ± 6.89 g/mL IC_{50}	PC-3 and Vero	50%	Touliabah et al. (2022)
Polycladia myrica	SeNPs	14.86 µg/mL and 50 mg/mL IC_{50}	HCT-116 and EAC cell line	50%	Abo-Neima et al. (2023)
Saccharina cichorioides	AgNPs	25 µg/mL	Rat C6 glioma cells	+	Yugay et al. (2020)

(continued)

Table 1.3 (continued)

Nanoparticles derived from marine plants	NPs	Inhibitory Dose	Cell lines	Inhibition activity	Reference
Sargassum glaucescens	AuNPs	4.75 ± 1.23, 7.14 ± 1.45, 10.32 ± 1.5, and 11.82 ± 0.9 µg/mL	HeLa, HepG2, CEM-ss, and MDA-MB-231 cell lines	50%	Ajdari et al. (2016)
Sargassum incisifolium	AgNPs	5.29 mM	MCF-7 and MCF-12a cell lines	40%	Mmola et al. (2016)
Sargassum incisifolium	AuNPs	4.17 mM	MCF-7 and HT-29 cell lines	40%	Mmola et al. (2016)
Sargassum longifolium	AgNPs	31 µg/mL	Hep-2 human laryngeal cell line	> 60%	Devi et al. (2013)
Sargassum muticum	AuNPs	23.83 ± 1.1 µg/mL, 18.75 ± 2.1 µg/mL, 12.5 ± 1.7 µg/mL, and 6.4 ± 2.3 µg/mL IC_{50}	HepG2 cells, MCF-7 cells, HeLa cells, and Jurkat cells	50%	Namvar et al. (2015a, b)
Sargassum muticum	AuNPs	4.22 ± 1.12, 5.71 ± 1.4, 6.55 ± 0.9, and 7.29 ± 1.7 µg/mL IC_{50}	K562, HL-60, Jurkat, and CEM-ss cells	50%	Namvar et al. (2015a, b)
Sargassum muticum	ZnONPs	150 µg/mL IC_{50}	Human liver cancer cell line (HepG2)	50%	Sanaeimehr et al. (2018)
Sargassum myriocystum	AgNPs	73.66 µg/mL IC_{50}	HeLa cells	50%	Balaraman et al. (2020)
Sargassum polycystum	AuNPs	300 mg mL^{-1} IC_{50}	MCF 7 breast cancer cell line	50%	Sivaraj et al. (2015)
Sargassum polycystum	AgNPs	20 µg/mL IC_{50}	Human colon cancer (HT-29) cells	50%	Palanisamy et al. (2017)

(continued)

1.9 Applications of Marine Flora-Derived Nanoparticles (MFNPs)

Table 1.3 (continued)

Nanoparticles derived from marine plants	NPs	Inhibitory Dose	Cell lines	Inhibition activity	Reference
Sargassum polycystum	AgNPs	135 mg mL^{-1} IC$_{50}$	MCF 7 breast cancer cell line	50%	Thangaraju et al. (2012)
Sargassum polycystum	CuONPs	100 µg mL^{-1}	MCF 7 breast cancer cell line	> 93%	Ramaswamy et al. (2016)
Sargassum swartzii	AuNPs	41.10 µg/mL IC$_{50}$	HeLa cells	50%	Dhas et al. (2014a, b)
Sargassum vulgare	AgNPs	2.84, 4.91, and 63.37 µg/mL CC$_{50}$	HL60, HeLa cells, and PBMC	50%	Govindaraju et al. (2015)
Sargassum wightii	MgONPs	37.5 ± 0.34 µg/mL IC$_{50}$	A549 lung cancer cell lines	50%	Pugazhendhi et al. (2019)
Spatoglossum schröederi	AgNPs	0.025 mg/mL	786-0 cells	20% MTT reduction rate	Amorim et al. (2016)
Turbinaria ornata	AgNPs	10.5 µg/mL IC$_{50}$	Y79 cell lines	50%	Remya et al. (2017)
Turbinaria turbinata	AgNPs	42.5 µg	Ehrlich Cell Carcinoma (ECC)		Bialy et al. (2017)
Turbinaria turbinata	AgNPs	98 µg/mL	Ehrlich Ascites Carcinoma (EAC)	99%	Khalifa et al. (2016)
Undaria pinnatifida	SeNPs	3.0 to 14.1 µM IC$_{50}$	A375 melanoma cells, HepG2 hepatocellular carcinoma cells, and MCF-7 breast adenocarcinoma cells	50%	Chen et al. (2008)
Undaria pinnatifida	FCD/LF NPs	20 µg/mL IC$_{50}$	PANC-1 cell line	50%	Etman et al. (2020)
Undaria pinnatifida	FCD/QC NPs	6.63 µg/mL IC$_{50}$	PANC-1 cell line	50%	Etman et al. (2021)
Red seaweed					

(continued)

Table 1.3 (continued)

Nanoparticles derived from marine plants	NPs	Inhibitory Dose	Cell lines	Inhibition activity	Reference
Marine red algae	AuNPs	12.5–400 µg/mL IC_{50}	HCT-116 human colon cancer cells, MDA-MB-231 human breast cancer cells, and HUVEC human umbilical vein endothelial cells	50%	Chen et al. (2018)
Marine red algae	CO_3O_4NPs	41.4 µg/mL IC_{50}	HepG2 cancer cells	50%	Ajarem et al. (2022)
Acanthophora spicifera	AuNPs	21.86 µg/mL IC_{50}	HT-29 cells	50%	Babu et al. (2020)
Amphiroa rigida	AgNPs	20 µg/mL IC_{50}	MCF-7 human breast cancer cell line	50%	Gopu et al. (2021)
Champia parvula	AgNPs	21.54 µg/mL and 42.36 µg/mL IC_{50}	Human lung (A549) and colon cancer (HT-29) cell lines	50%	Viswanathan et al. (2023)
Corallina elongata	AgNPs	98 µg/mL	Ehrlich Ascites Carcinoma (EAC)	74%	Khalifa et al. (2016)
Corallina officinalis	AuNPs	1.5 µL/mL	MCF-7 human breast cancer cell line	50%	El-Kassas and El-Sheekh (2014)
Gelidiella sp.	AgNPs	31.25 µg/mL IC_{50}	Hep-2 cell lines	50%	Devi et al. (2012)
Gelidium amansii	AuNPs	100 ppm	Rat C6 cell line	75%	Kumar et al. (2017a, b)
Gelidium corneum	AuNPs	50 mg/mL	A549, THP-1, and HL-60 cell lines		González-Ballesteros et al. (2022)
Gelidium crinale	AgNPs	98 µg/mL	Ehrlich Ascites Carcinoma (EAC)	81%	Khalifa et al. (2016)

(continued)

1.9 Applications of Marine Flora-Derived Nanoparticles (MFNPs)

Table 1.3 (continued)

Nanoparticles derived from marine plants	NPs	Inhibitory Dose	Cell lines	Inhibition activity	Reference
Gelidium pusillum	AgNPs	43.09 ± 1.6 μgmL^{-1} IC$_{50}$	MDA-MB-231 cancerous cells	50%	Jeyarani et al. (2020)
Gracilaria corticata	AgNPs	62.5 μg/mL IC$_{50}$	MCF7 cells	50%	Bhimba et al. (2015)
Gracilaria corticata	AgNPs	42.5 μg/mL IC$_{50}$	MCF-7 cell line	50%	Parthasarathy et al. (2021)
Gracilaria edulis	AgNPs and ZnONPs	28.55 μg/mL IC$_{50}$ and 39.60 μg/mL IC$_{50}$	PC3 human prostate cancer cell lines	62–70%	Priyadharshini et al. (2014)
Gracilaria edulis	ZnONPs	35 ± 0.03 μg/mL	SiHa cells	50%	Asik et al. (2019)
Gracilaria foliifera	AgNPs	30 μg mL^{-1} IC$_{50}$	HFb-4 and MCF-7 cell lines	50%	Algotiml et al. (2022b)
Gracilaria foliifera	AgNPs	10 μg mL^{-1} LC$_{50}$	A. salina nauplii	??	Algotiml et al. (2022b)
Halymenia poryphyroides	AgNPs	200 μg/mL	DLA and EAC cell lines	$89.36 \pm 3.12\%$ and $91.45 \pm 4.72\%$	Vishnu and Murugesan (2014)
Halymenia pseudofloresii	AuNPs	19.02 μg/mL and 32.46 μg/mL IC$_{50}$	A549 lung cancer and LN-18 glioblastoma cancer cells	50%	Palaniyandi et al. (2023)
Halymenia venusta	AuNPs	39.04 μg/mL IC$_{50}$	A549 cells	50%	Baskar et al. (2023)
Hypnea valentiae	AgNPs	24.6 μg/mL and 5.917 μg/mL IC$_{50}$	HT-29 human colon cancer and A549 lung cancer cell lines	50%	Viswanathan et al. (2022)
Kappaphycus	EPI-CAO-AuNPs	0.087 ± 0.036 μmol/L IC$_{50}$	HepG2 cells	50%	Chen et al. (2019)

(continued)

Table 1.3 (continued)

Nanoparticles derived from marine plants	NPs	Inhibitory Dose	Cell lines	Inhibition activity	Reference
Laminaria japonica	Doxorubicin (DOX) loaded Protamine/fucoidan NPs	2.5 μg/mL	MDA-MB-231 cells	44.5%	Lu et al. (2017)
Laurencia obtusa	AgNPs	98 μg/mL	Ehrlich Ascites Carcinoma (EAC)	86%	Khalifa et al. (2016)
Polysiphonia sp.	AgNPs	100 μg/mL	MCF-7 cell lines	95.5%	Moshfegh et al. (2019)
Porphyra linearis	AuNPs	4 mg/mL	A549, THP-1, and HL-60 cell lines		González-Ballesteros et al. (2022)
Porphyra vietnamensis	AuNPs	15 μg/mL	LN-229 human glioma cell line	80%	Venkatpurwar et al. (2011)
Porphyra vietnamensis	AuNPs	10 μM	Vero Cells	98% ± 2%	Venkatpurwar et al. (2012)
Portieria hornemannii	AgNPs	–	Vero Cells	75.4%	Sabatini and Anchana Devi (2017)
Pterocladia capillacea	CuONPs	0.40 ± 0.08, 1.50 ± 0.12, and 0.70 ± 0.09 μg/mL IC_{50}	Hepatocellular carcinoma (HEP-G2), breast cancer (MCF-7) and ovarian (SKOV-3) cancer cell lines	50%	Aboeita et al. (2022)
Pterocladiella capillacea	AgNPs	3.7 μg/mL IC_{50}	HepG2	50%	Kassas and Attia (2014)
Rosenvingea intricata	Fe_3O_4NPs	311.7 & 460.5 μg/mL IC_{50}	Hep3B and PANC1 cells	50%	Sri et al. (2024)
Spyridia filamentosa	AgNPs	100 μg mL^{-1}	MCF-7 cells	Strong toxic activity	Valarmathi et al. (2020)
Seagrasses					

(continued)

1.9 Applications of Marine Flora-Derived Nanoparticles (MFNPs)

Table 1.3 (continued)

Nanoparticles derived from marine plants	NPs	Inhibitory Dose	Cell lines	Inhibition activity	Reference
Cymodocea serrulata	AgNPs	34.5 and 61.24 μg mL^{-1} GI$_{50}$	HeLa cervical cancer and African green monkey kidney cells (Vero)	50%	Chanthini et al. (2015)
Cymodocea serrulata	AgNPs	100 μg/mL LD$_{50}$	Human lung cancer A549 cells	50%	Palaniappan et al. (2015)
Mangroves					
Avicennia marina	AgNPs	50 μg/mL	A549 lung cancer cells	54%	Tian et al. (2020)
Excoecaria agallocha	AgNPs	100 μg/mL IC$_{50}$	MCF-7 human breast cancer cell line	92%	Bhuvaneswari et al. (2017)
Rhizophora apiculata	AgNPs	50 μg/mL	A375 (Skin Cancer), A549 (Lung Cancer), and KB-3-1 (Oral Cancer)	68.16%, 43.91%, and 77.41%	Alsareii et al. (2022)
Sonneratia alba	AgNPs	100 μg/mL	Vero cells	35%	Murugan et al. (2017)
Sea-blites and other plants associated with marine plants					
Citrullus colosynthis	AgNPs	500 nM	HEp 2 cells	50%	Satyavani et al. (2011)
Ipomoea pes-caprae	AgNPs	75 μg/mL IC$_{50}$	MCF-7 cancer cell lines	50%	Veeramani et al. (2018)
Suaeda monoica	AgNPs	500 nm μg/mL IC$_{50}$	Hep-2 cell line	50%	Satyavani et al. (2012)

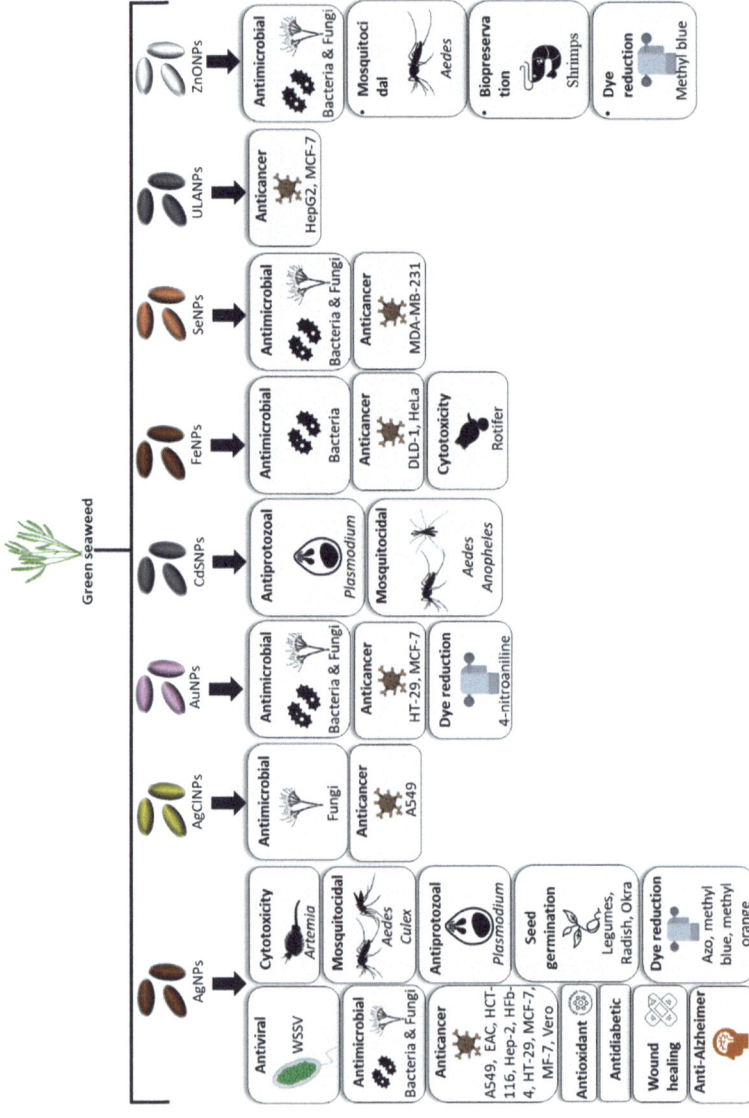

Fig. 1.8 Illustration depicting the applications of various nanoparticles derived from green seaweeds

1.9 Applications of Marine Flora-Derived Nanoparticles (MFNPs)

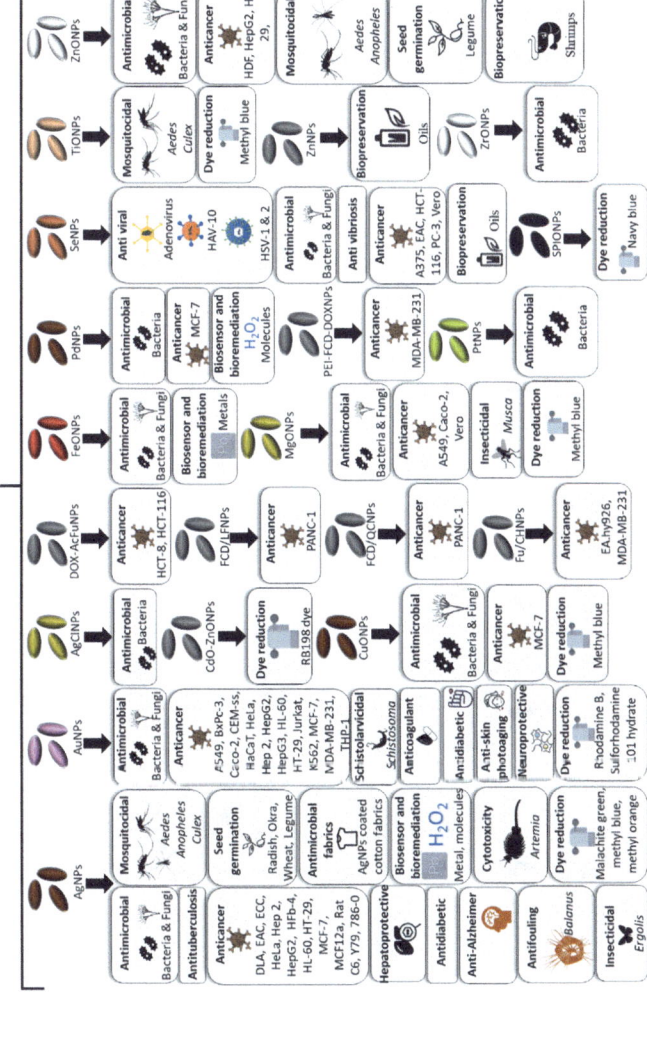

Fig. 1.9 Illustration depicting the applications of various nanoparticles derived from brown seaweeds

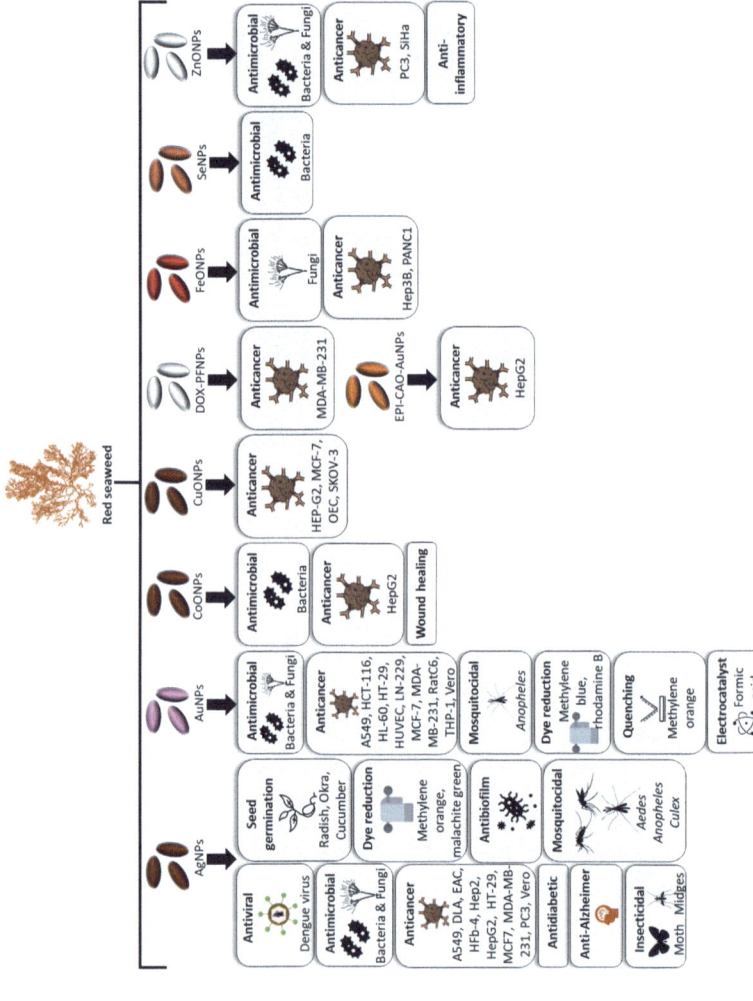

Fig. 1.10 Illustration depicting the applications of various nanoparticles derived from red seaweeds

1.9 Applications of Marine Flora-Derived Nanoparticles (MFNPs)

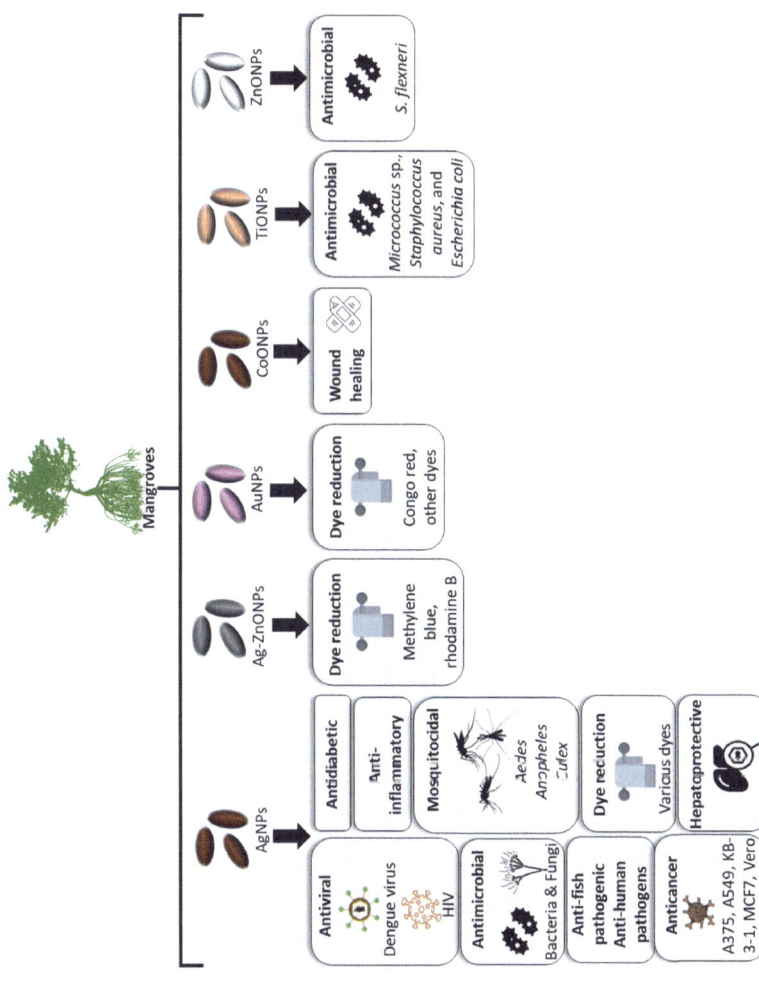

Fig. 1.11 Illustration depicting the applications of various nanoparticles derived from mangroves

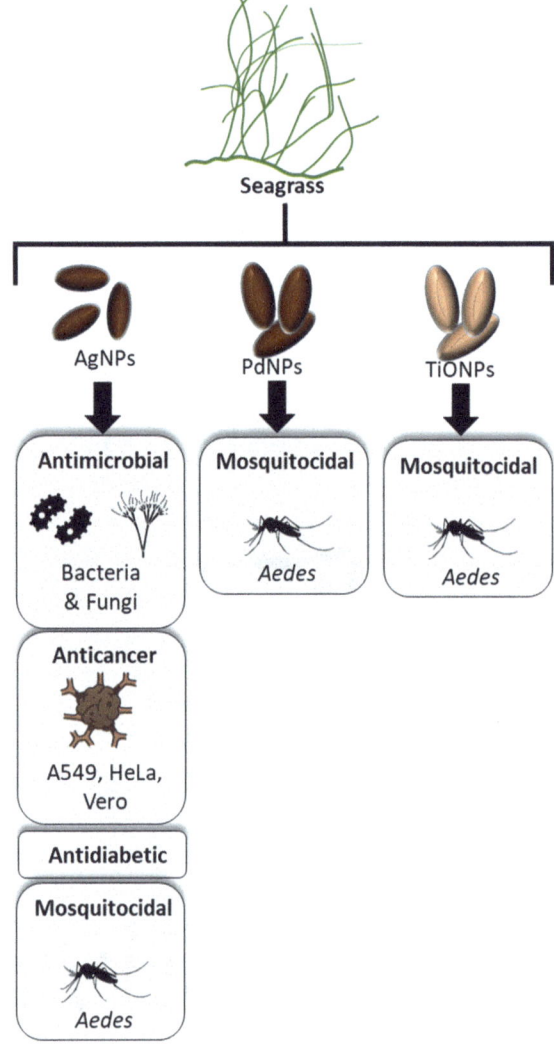

Fig. 1.12 Illustration depicting the applications of various nanoparticles derived from seagrasses and other coastal plants

Sargassum muticum, with concentrations of 4.22 ± 1.12, 5.71 ± 1.4, 6.55 ± 0.9, and 7.29 ± 1.7 μg/mL IC_{50} inhibited K562, HL-60, Jurkat, and CEM-ss cells, respectively (Namvar et al. 2015b). *Sargassum polycystum* derived AgNPs did inhibited 50% of HT-29 cells at 20 μg/mL IC_{50} (Palanisamy et al. 2017). AgNPs synthesized from *Turbinaria ornata* at 10.5 μg/mL IC_{50} displayed 50% of inhibition of Y79 cell lines (Remya et al. 2017).

Among red seaweeds, *Amphiroa rigida* derived AgNPs displayed 50% inhibition of MCF-7 cell lines at 20 μg/mL IC_{50} (Gopu et al. 2021). *Corallina officinalis* derived AuNPs with 1.5 μL/mL concentration caused 50% inhibition of MCF-7

1.9 Applications of Marine Flora-Derived Nanoparticles (MFNPs)

Fig. 1.13 Illustration depicting the applications of various nanoparticles derived from other coastal plants

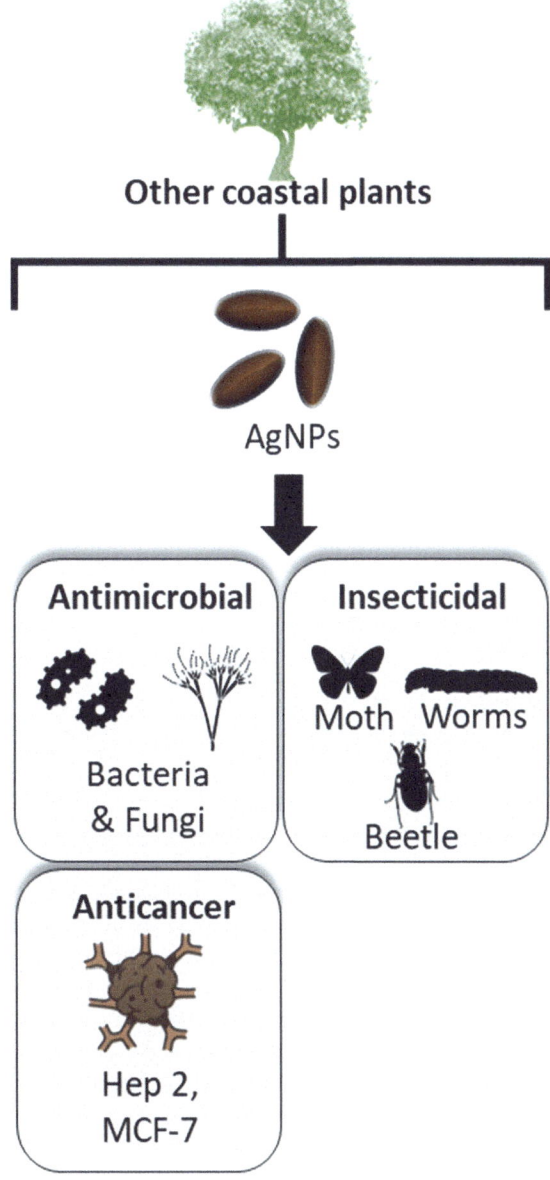

cell lines (El-Kassas and El-Sheekh 2014). AgNPs synthesized from *Gracilaria edulis* demonstrated 50% inhibitory activity against PC3 human prostate cancer cell lines at 28.55 μg/mL IC$_{50}$ (Priyadharshini et al. 2014). *Halymenia pseudofloresii* derived AuNPs at 19.02 μg/mL IC$_{50}$ showed 50% anticancer activity against A549 lung cancer cell lines (Palaniyandi et al. 2023). *Porphyra vietnamensis* synthesized

AuNPs at 15 μg/mL showed 80% anticancer activity against LN-229 human glioma cell lines (Venkatpurwar et al. 2011). CuONPs fabricated from *Pterocladia capillacea* demonstrated 50% of inhibition against HEP-G2, MCF-7 and SKOV-3 cancer cell lines at 0.40 ± 0.08, 1.50 ± 0.12, and 0.70 ± 0.09 μg/mL IC$_{50}$ (Aboeita et al. 2022).

Seagrass *Cymodocea serrulata* derived AgNPs showed potential anticancer activity against HeLa and Vero cells at 34.5 and 61.24 μg/mL GI$_{50}$ dosages, correspondingly (Chanthini et al. 2015). Among mangroves, *Avicennia marina* derived AgNPs demonstrated 50% inhibition of A549 cells at 54 μg/mL concentration (Tian et al. 2020). *Rhizophora apiculata* derived AgNPs at 50 μg/mL concentration inhibited A375, A549, and KB-3–1 cells with 68.16%, 43.91%, and 77.41% inhibition, respectively (Alsareii et al. 2022). Other coastal plants like *Ipomoea pes-caprae* derived AgNPs displayed 50% inhibition of MCF-7 cells at 75 μg/mL IC$_{50}$ dosage (Veeramani et al. 2018).

1.9.6 Antidiabetic Activity

AgNPs fabricated from *Colpomenia sinuosa* inhibited α-amylase and α-glucosidase with an IC$_{50}$ value of 490 ± 0.02 mg/mL and 385 ± 0.02 mg/mL, correspondingly, compared to standard drug acarbose (Kiran and Murugesan 2014a). Silver nanoparticles synthesized from *Galaxaura elongata, Turbinaria ornata,* and *Enteromorpha flexuosa* showed promising amylase inhibitory action of 64.20%, 67.46% and 67.46%, respectively, at 200 μg/mL concentration, while standard acarbose inhibitory action was 73.25% (Azeem et al. 2022). Antidiabetic activity of AgNPs synthesized from *Portieria hornemannii* extract inhibited 53.1% of amylase activity at 0.4. mL concentration (Sabatini and Anchana Devi 2017).

AgNPs green synthesized from the aqueous extract of seagrass, *Enhalus acoroides*, showed antidiabetic activity by inhibiting α-glucosidase enzyme at IC$_{50}$ 47 μg/mL concentration (Senthilkumar et al. 2016).

The green synthesized AgNPs obtained from the bark extracts of *X. granatum* and leaf extract of *A. officinalis* have shown antidiabetic properties by inhibiting α-amylase (IC$_{50}$ values of 0.19 mg/mL and 0.28 mg/mL) and α-glucosidase enzymes (IC$_{50}$ values of 0.13 and 0.15 mg/mL), respectively (Das et al. 2019).

1.9.7 Wound Healing Properties

Silver nanoparticles (40 μg/mL) synthesized from *C. scalpelliformis* leave extract showed wound healing property in albino rats (Manikandan et al. 2019). Cobalt oxide nanoparticles derived from red algae demonstrated anticoagulant and thrombolytic properties to human blood cells (Ajarem et al. 2022).

Green synthesized AgNPs originated from *Rhizophora apiculata* leaf extract reported to have 82.79% of wound healing property as evidenced with L929 cell lines (Alsareii et al. 2022).

1.9.8 Anti-inflammatory Properties

ZnONPs synthesized using Kappa-Carrageenan have demonstrated 82% anti-inflammatory activity at a concentration of 500 μg/mL (Vijayakumar et al. 2020). AgNPs green synthesized from the leaf extract of *Rhizophora apiculata* showed anti-inflammatory activity (71.65 ± 0.88%) at 500 μg/mL concentration (Alsareii et al. 2022).

1.9.9 Anti-Alzheimer Activity

The biogenic AgNPs synthesized from *Galaxaura elongata, Turbinaria ornata,* and *Enteromorpha flexuosa* inhibited AChE proteins at IC_{50} concentration of 71.92 μg/mL, 76.14 μg/mL, and 19.78 μg/mL, with inhibition action of 73.75%, 70.53% and 63.46%, respectively; while standard donepezil displayed 83.6% inhibition at IC_{50} 71.4 μg/mL (Azeem et al. 2022).

1.9.10 Antiprotozoal/Antiplasmodial Activity

CdS nanoparticles derived from *Valoniopsis pachynema* extract showed antiplasmodial activity against chloroquine-resistant (CQ-r) and CQ-sensitive (CQ-s) strains of *Plasmodium falciparum* with IC_{50} values of 89.21 μg/mL (CQ-r) and 76.14 μg/mL (CQ-s), respectively (Sujitha et al. 2017). AgNPs synthesized from *U. lactuca* extracts demonstrated higher antiplasmodial activity by inhibiting *P. falciparum* at a concentration of IC_{50} values 76.33 μg/mL (CQ-s) and 79.13 μg/mL (CQ-r) compared to chloroquine (Murugan et al. 2015b). Similarly, *C. tomentosum* synthesized AgNPs inhibited CQ-s and CQ-r strains of *P. falciparum* at IC_{50} of 72.45 μg/mL and 76.08 μg/mL, respectively (Murugan et al. 2016b).

1.9.11 Antifouling Activity Nanoparticles

Facile green AgNPs fabricated from *Turbinaria ornata* displayed 100% mortality of fouling organism *Balanus amphitrite* larvae within 24 h at 250 μg mL^{-1} concentration (Krishnan et al. 2015).

1.9.12 Mosquitocidal Activity

AgNPs synthesized from *C. scalpelliformis* displayed mosquitocidal activity against first-instar to fourth instar and pupa of vector *Culex quinquefasciatus* at LC_{50} of 3.08 ppm, 3.49 ppm, 4.64 ppm, 5.86 ppm, and 7.33 ppm, respectively (Murugan et al. 2015a). Similarly, AgNPs derived from *U. lactuca* inhibited the larval forms of *A. stephensi* with LC_{50} concentrations ranging between 18.365 ppm to 43.012 ppm (Murugan et al. 2015b). *Ulva lactuca* fabricated AgNPs inhibited 95% of *Ae. aegypti* and *Cu. pipiens* at IC_{95} concentrations of 80.51 ppm and 105.65 ppm (Aziz 2022). The IC_{95} value against *Ae. aegypti* was 226.9 ppm and for *Cu. pipiens* was 337.19 ppm (Aziz 2022). CdS nanoparticles synthesized from *Valoniopsis pachynema* demonstrated mosquitocidal activity against larva I of *A. stephensi* (LC_{50} 16.856 µg/mL) and *An. sundaicus* (LC_{50} 13.584 µg/mL) (Sujitha et al. 2017). Zinc oxide nanoparticles derived from *Ulva lactuca* caused 100% mortality of fourth instar larvae of *Aedes aegypti* within 24 h at LC_{50} 22.38 µg/mL concentration (Ishwarya et al. 2018a).

Ovideterrence assays revealed that AgNPs (30 ppm) fabricated from *Sargassum muticum* reduced 100% egg hatchability rates in *A. aegypti, A. stephensi*, and *C. quinquefasciatus* (Madhiyazhagan et al. 2015). Mosquitocidal activity of AgNPs synthesized from *S. muticum* was effective against adult Indian strains of *Cx. quinquefasciatus, Ae. aegypti* and *An. stephensi* at LC_{50} concentrations of 42.3, 34.3 and 29.7 mg/mL, respectively, as well as against Saudi Arabian strains, *Ae. aegypti* and *Cx. pipiens* with LC_{50} concentrations of 86.4 and 120.0 mg/mL, respectively (Trivedi et al. 2021). AgNPs fabricated from *Sargassum myriocystum* (Balaraman et al. 2020) and *Sargassum polycystum* (Vinoth et al. 2019) demonstrated 100% and 90% mortality of *Ae. Aegypti* and *Cx. quinquefasciatus* at LC_{50} concentrations of 6.90 mg/L and 5.59 mg/L, and LC_{90} concentrations of 85.81 and 178.85, respectively. AgNPs synthesized from *Turbinaria ornata* showed larvicidal activity against *Aedes aegypti, Anopheles stephensi* and *Culex quinquefasciatus* at LC_{50} of 0.738, 1.134, and 1.494 µg/mL, correspondingly (Deepak et al. 2017). AgNP synthesized from *C. tomentosum* displayed higher toxicity on *A. stephensi* pupae with LC_{50} value of 040.7 ppm (Murugan et al. 2016b). TiO_2 NPs fabricated from *Sargassum myriocystum* showed potential larvicidal activity against *Aedes aegypti* and *Culex quinquefasciatus* (Balaraman et al. 2022). ZnONPs synthesized from *S. wightii* inhibited (50%) third instar larvae of *Ae. aegypti* (Ishwarya et al. 2018b) and *Anopheles stephensi* (Murugan et al. 2018) mosquitoes at $LC_{50} = 49.22$ mg/L and $LC_{50} = 5.696$ ppm, respectively. Murugan et al. (2018) observed an increase in the larval predation efficiency of *Poecilia reticulata* upon post treatment with ZnONPs (Murugan et al. 2018).

Green synthesized AgNPs obtained from aqueous extract of red seaweed, *Amphiroa rigida*, showed 100% mortality of 3rd and 4th instar larvae of *Aedes aegypti* at 20 and 40 µg concentrations, respectively (Gopu et al. 2021). AgNPs synthesized from *Centroceras clavulatum* extracts displayed mosquitocidal activity against *A. aegypti* IV larvae (LC_{50} 29.155 ppm) and pupa (LC_{50} 33.877 ppm) (Murugan et al. 2016a). *Gracilaria corticata* derived AgNPs have demonstrated larvicidal activity at a concentration of 100 µg/mL within 12 h against *Ae. aegypti* ($LC_{50} = 133.0$), *An.*

stephensi ($LC_{50} = 88.8$), and *Cx. quinquefasciatus* ($LC_{50} = 88.4$) (Naveenkumar et al. 2023). *Gracilaria edulis* derived AgNPs demonstrated 100% larvicidal activity against *Cx. quinquefasciatus* at 17 ppm of LC_{50} concentration (Madhiyazhagan et al. 2017). *Hypnea musciformis* originated AgNPs displayed larvicidal (LC_{50} value of 18.14–38.23 ppm against 1st instar larvae and pupae, respectively) and pupicidal (LC_{50} value of 24.5–38.23 ppm against 1st instar larvae and pupae, respectively) activity against the dengue vector *Aedes aegypti* (Roni et al. 2015). Similarly, AgNPs fabricated from *Gracilaria firma* demonstrated mosquitocidal against *A. aegypti* instar I (LC_{50} 25.895) and pupae to (LC_{50} 351.419) (Kalimuthu et al. 2017). AgNPs synthesized from *Portieria hornemannii* demonstrated 100% inhibition of *Aedes aegypti* eggs and larvae at 8 µL concentration (Sabatini and Anchana Devi 2017). *Gracilaria crassa* extracts derived AuNPs inhibited I to IV instars of *Anopheles stephensi* larvae at LC_{50} concentration range of 69.34 ppm –79.81 ppm (Kamaraj et al. 2022).

Seagrass, *Cymodocea serrulata* derived AgNPs, PdNPs, and TiO_2NPs have demonstrated potential mosquito larvicidal against 1st instar larvae of *A. stephensi*, at a concentration of $LC_{50} = 2.099$ µg/mL, $LC_{50} = 4.010$ µg/mL, and $LC_{50} = 5.052$ µg/mL, correspondingly (Amutha et al. 2019).

AgNPs biosynthesized from the leaf aqueous extracts of mangrove, *Avicennia marina*, has demonstrated in vitro larvicidal activity at LC_{50} value of 4.374 and 7.406 mg/L against *Anopheleus stephensi* and *Aedes aegypti* larvae, respectively (Balakrishnan et al. 2016). *Acanthus ilicifolius* leaf extract derived AgNPs displayed potential larvicidal activity against *A. subalbatus* and *Ae. aegypti* at LC_{90} concentrations of 2.13 and 5.98 mg/L, correspondingly (Ali et al. 2015). Similarly, larvicidal activity of AgNPs fabricated from *A. marina* was observed at $IC_{50} = 16.5586$ ppm against fourth instar larvae of dengue fever mosquito *Aedes aegypti* (Barnawi et al. 2019). Mangrove *Excoecaria agallocha* derived AgNPs showed highest inhibition activity against 3rd and 4th instar larvae of *Aedes aegypti* at LC_{50} concentrations of 4.65 mg/L and 6.10 mg/L (Kumar et al. 2016). Silver nanoparticles synthesized from leaf extracts of *Rhizophora mucronata* showed larvicidal activity against *Aedes aegypti* and *Culex quinquefasciatus* with an inhibition concentration (LC_{50}) of 0.585 and 0.891 mg/L, respectively. *Sonneratia albu*-synthesized AgNPs killed *Aedes aegypti* larva I and pupa with LC_{50} concentration of 3.15 (I) and 13.61 ppm, correspondingly (Murugan et al. 2017).

1.9.13 Insecticidal Activity

MgO-NPs synthesized from brown Algae, *Cystoseira crinita* caused mortality of first instar lave ($99.0 \pm 1.22\%$) and pupa ($81.0 \pm 3.16\%$) of *Musca domestica* with LC_{50} concentration of 3.08 and 5.86 (Fouda et al. 2022b). Silver nanoparticles extracted from *Sargassum muticum* showed insecticidal activity against fourth instar *Ergolis merione* at LC_{50} of 4.5 µL (Moorthi et al. 2015).

Hypnea musciformis originated AgNPs displayed inhibition of cabbage pest *Plutella xylostella* (Roni et al. 2015). *Gracilaria edulis* derived AgNPs demonstrated 100% larvicidal activity against *Chironomus circumdatus* at 29 ppm LC_{50} concentration (Madhiyazhagan et al. 2017).

AgNPs derived from *Hibiscus tiliaceus* extracts demonstrated potential anti pest activity against *T. casutaneum*, *R. dominica*, *Sitophilus oryzae*, with mean toxicity values of 45.2 ± 0.24, 42.8 ± 0.24, and 37.4 ± 0.16 at 150 µg/cm^2 (Usha Rani et al. 2016). Also, antifeedant activity also reported for AgNPs against *S. litura* and *H. armigera* (Usha Rani et al. 2016).

1.9.14 Seed Germination Properties

The seeds of *Raphanus sativus* var. *longipinnatus* when treated with AgNPs synthesized from the aqueous extracts of *Halimeda gracilis* displayed 100% seed germination and root elongation properties with high index values (Roy and Anantharaman 2018b). AgNPs fabricated from *C. antennina* (Kingslin et al. 2022b) and *Enteromorpha prolifera* (Kingslin et al. 2022a) have also showed 60 to 100% germination property with seeds of *Vigna unguiculata*, *Vigna radiata*, and *Cicer arietinum*. AgNPs synthesized from *Chaetomorpha antennina* (Roy and Anantharaman 2017) and *Sargassum ilicifolium* (Suparna Roy and Anantharaman 2018a, b, c, d), have also showed seed germination efficiency (60–100%) in the seeds of *Abelmoschus esculentus* and *Raphanus sativus* var. *longinnatus*.

Laminaria digitata-synthesized ZnONPs showed dose dependent root and shoot germination efficacy in mung *Vigna radiata* (Vijayakumar et al. 2022). *Laminaria japonica* derived AgNPs, at all concentrations (10, 20, 30, 40, 60, and 80 ppm), demonstrated 90 to 98% seed germination ability in both *T. aestivum* and *P. mungo* seeds (Kim et al. 2018). Species specific concentration of AgNPs (100%, 75%, and 25%) showed 100%, 90% and 90% seed germination effect on *Vigna unguiculata*, *Vigna radiata*, and *Cicer arietinum* (Kingslin and Ravikumar 2016).

AgNPs synthesized from the red seaweed *Amphiroa anceps* when applied to the seeds of *Raphanus sativus* var. *longipinnatus* and *Abelmoschus esculentus* displayed highest seed germination ability (80% and 75%, respectively) compared to seeds treated with normal water (20–40%) (S. Roy and Anantharaman 2018c). Silver nanoparticles fabricated from *P. hornemannii* demonstrated seed germination activity in cucumber, *Cucumis sativus*, by showing maximum growth within 7 days of incubation at 5 µg/mL concentration (Aravindan et al. 2014).

1.9.15 Biopreservation Properties

Zinc oxide nanoparticles fabricated from *U. fasciata* extracts showed biopreservation property by reducing bacterial load on refrigerated shrimps at 4 °C and maintained acceptable sensorial attributes than controls (Alsaggaf et al. 2021). Zinc and selenium nanoparticles derived from *Sargassum latifolium* were inferred that these nanoparticles can be employed to preserve unstable edible oils like corn and soybean oils from oxidation (El-Khateeb et al. 2019).

1.9.16 Applications in Textile Industry

An accelerated azo dye reduction by AgNPs fabricated from *Caulerpa serrulata* was reported in presence of increased concentration of $NaBH_4$ or catalytic dosage (Aboelfetoh et al. 2017). AgNPs synthesized from *Caulerpa racemosa* (Edison et al. 2016) and *Ulva lactuca* (Kumar et al. 2013a) demonstrated high catalytic activity toward the degradation of methyl blue and methyl orange dye, respectively. ZnONPs (200 mg) as catalyst showed photodegradation of methyl blue under sunlight (Ishwarya et al. 2018a). Likewise, ZnONPs (1.0 mg mL^{-1}) fabricated from *U. fasciata* showed $84.9 \pm 1.2\%$ dye removal as well as decolorized tanning wastewater up to $96.1 \pm 1.7\%$ (Fouda et al. 2022a). Reduction of 4-nitroaniline by potassium borohydride in presence of AuNPs obtained from *Ulva faciata* was evident with the kinetic constant of 1.55×10^{-2} min^{-1} (Kumari et al. 2014).

Facile green synthesized MgONPs from *Sargassum wightii* aqueous extract showed high photocatalytic activity for methylene blue degradation (Pugazhendhi et al. 2019). *Turbinaria decurrens* extract derived SPIONPs showed maximum absorption capacity of 5.353 mg g^{-1} for navy blue dye (Khaleelullah et al. 2017). Similarly, CuONPs fabricated from *Cystoseira trinodis* showed 89% degradation of methylene blue at optimum pH 4 (Gu et al. 2018). AuNPs synthesized from brown algae have proven to act as catalyst for the reduction of organic dye molecules Rhodamine B and Sulforhodamine 101 hydrate (Ramakrishna et al. 2016). *Padina gymnospora* extracts derived CdO-ZnONPs degraded 99.57% percentage of RB198 dye under sunlight within 15 min (Rajaboopathi and Thambidurai 2017). AgNPs synthesized from *Sargassum ilicifolium* demonstrated degradation of malachite green (82.9%) and methylene blue (100%) in aqueous medium under light exposure (Devi et al. 2022). AgNPs (Balaraman et al. 2020) and TiO_2NPs (Balaraman et al. 2022) synthesized from *Sargassum myriocystum* demonstrated 98% and 92.92% degradation of methylene blue within 60 min and 45 min, respectively. *Sargassum wightii* and *Hypnea musciformis* derived AgNPs did show similar dye degradation property when 50 mL of methyl orange solution was treated with 20 mg of AgNPs (Selvam and Sivakumar 2015, 2014).

Cotton fabrics treated with 2% citric acid and 1% Binder when treated with AgNPs (108 ppm concentration) derived from crude hot water-soluble polysaccharides of green, brown, and red macroalgae displayed 100% antimicrobial activity against *S. aureus* and *E. coli* (El-Rafie et al. 2013).

Under sunlight exposure, AuNPs synthesized from *Gelidiella acerosa* degraded commercial dyes methylene blue and rhodamine B (Subbulakshmi et al. 2022). The degradation of malachite green by AgNPs synthesized from *Gracilaria corticata* was evidenced within 6 h of incubation time via reduction in peak intensity (Poornima and Valivittan, 2017).

Green Ag-ZnO nanoparticles fabricated from *Excoecaria agallocha* leaf extract showed 98.83% degradation of methylene blue and rhodamine B dyes via photocatalytic activity under solar irradiation (Khan et al. 2020). AuNPs green synthesized from tannin rich extract obtained from the bark of *Xylocarpus granatum* showed a heterogeneous catalyst property by degrading Congo Red from the AuNPs coated fabrics (Pisitsak et al. 2021). Similarly, AgNPs synthesized from polysaccharides extract of *Xylocarpus granatum* showed considerable decolorization rate (Maity and Mondal, 2017). Further, AgNPs and AuNPs synthesized from numerous marine plants have the dye removal property (Devasena and Thiruchelvi 2019). These studies suggest that NPs have an important application in treating dye effluents in waste water and industrial water before being released into natural environment.

1.9.17 Biosensor and Bioremediation Properties

Palladium nanoparticles synthesized from *Sargassum bovinum*, have the biosensor property and stability to detect H_2O_2 in the range of 5.0 µM–15.0 mM (Momeni and Nabipour, 2015). Silver nanoparticles synthesized from different algae have been reviewed to have biosensor application to detect various different metal pollutants (Chugh et al. 2021). AgNPs synthesized from deoiled *Saccharina japonica* powder (DSP) suggested to act as biosensor to detect H_2O_2 from various samples by showing decreased absorption to various concentrations of H_2O_2 (Sivagnanam et al. 2017).

Iron oxide nanoparticles (Fe_3O_4-NPs) synthesized from *P. pavonica* and *S. acinarium* displayed bioremediation property by high removal of Pb 91% and 78%, respectively, upon entrapping them in calcium alginate beads (El-Kassas et al. 2016).

1.9.18 Fluorescence Enhancement or Quenching Application

Gold nanoparticles synthesized from red seaweed *Osmundaria obtusiloba* enhanced the fluorescence signals of fluorophore, methyl orange (Rojas-Pérez et al. 2015), indicating their importance as fluorescence quenchers in photoluminescence studies. A study reported the synthesis and biological properties of fluorescent AuNPs synthesized from freshwater red alga *Lemanea fluviatilis* (Sharma et al. 2014).

1.9 Applications of Marine Flora-Derived Nanoparticles (MFNPs)

1.9.19 Electrocatalyst Properties

Gold nanoparticles derived from *Actinotrichia fragilis* displayed excellent electrocatalyst activity for formic acid oxidation without CO poisoning (Momeni et al. 2016).

1.9.20 Other Biological Properties

The green synthesized FeNPs obtained from *U. flexuosa* showed low toxicity to brackish water rotifer *B. rotundiformis* at EC_{50} > 1000 mg/L concentration (Mashjoor et al. 2018). AgNPs synthesized from the aqueous extract of *Ulva fasciata* demonstrated antioxidant and hepatoprotective properties against liver tissues induced by CCl_4 toxicity (Alshubaily et al. 2020). AgNP nanoparticles fabricated from aqueous extract of *Ulva rigida* showed 50% cytotoxicity to brine shrimp *Artemia salina* larvae at LC_{50} 4.5 µg/mL concentration (El-Kassas and ElKomi 2014).

AgNP nanoparticles fabricated from aqueous extract of *Dictyota bartayresiana* (Antonysamy et al. 2015), *Sargassum ilicifolium* (Kumar et al. 2012a), and *Turbinaria conoides* (Vijayan et al. 2014), showed 50% cytotoxicity to brine shrimp *Artemia salina* larvae at LC_{50} 196.5µL/l, LD_{50} 10 nM/mL, and LC_{50} 88.14 ± 5.04 µL/mL concentrations, respectively. Au-NPs synthesized from *Cystoseira myrica* displayed schistolarvicidal activity (64.2%) against *Schistosoma mansoni* (Kamal et al. 2022). Hepatoprotective property has been reported from the AgNPs fabricated from *Sargassum siliquosum* (Vasquez et al. 2016). Significant antidiabetic activity (57.3%) and anticoagulant activity (73.31 ± 0.89%) of AuNPs synthesized from *Turbinaria conoides* was observed at 200 µg/mL concentration (Venkatraman et al. 2018). Gold nanoparticles synthesized from *Ecklonia stolonifera* thought to have properties to treat skin-photoaging (Jun et al. 2020). *Dictyopteris divaricata* extract's synthesized AuNPs showed promising neuroprotective properties (Park et al. 2019).

Gelidiella acerosa derived AgNPs demonstrated quorum sensing inhibition to stop violaccin production and also antibiofilm activity against *Vibrio* species (Satish et al. 2017). The incorporation of metal oxide nanoparticles (ZnO, CuO, and SiO2) into κ-Carrageenan/PEG 3000 biofilms significantly enhanced their mechanical strength, optical clarity, and water barrier functionality (Sudhakar et al. 2022).

In addition to the known antimicrobial activity of AgNPs synthesized from aqueous leaf extract of *Rhizophora apiculata* (Antony et al. 2011), hepatoprotective effect has also been reported (Zhang et al. 2019).

1.10 Enhancing Biological Properties of Nanoparticles by Conjugation

The stabilization and the enhanced biological properties of silver nanoparticles are reported to be much higher when conjugated with other bioactive phytochemicals (Ram et al. 2022) or antibiotics/drugs (Lu et al. 2017; Pawar et al. 2019; Rajeshkumar 2017). A study found that adding nedaplatin to CuONPs derived from red algae has yielded potential anticancer properties (Aboeita et al. 2022), and similarly enhanced antimicrobial activity was reported against disease causing bacteria treated with polyvinylpyrrolidone and PtNP composite (V. Sri Ramkumar et al. 2017a, b). These reports suggests that the enhanced bioactive efficacy and targeted delivery of marine plant derived nanoparticles can be improved by conjugation/combining of NPs with desired marine natural extract/bioactive compound. Conjugation step potentially improves the biological property of conjugate (nanoparticle + bioactive compound), which would help to treat multidrug resistant microbial infections and various tumor cells. Also, employing crude extracts like crude hot water soluble polysaccharide (El-Rafie et al. 2013) or pure compounds such as ulvan (Massironi et al. 2019), alginic acid (Francavilla et al. 2014; Parker et al. 2015), agar (Francavilla et al. 2014), carrageenan (Alvarez-Vinas et al. 2022), epirubicin-loaded kappa-carrageenan (Chen et al. 2019), carrageenan oligosaccharide (Chen et al. 2018), starch (Francavilla et al. 2014), fucoidan (Etman et al. 2021; Khan et al. 2019; Soisuwan et al. 2010), fucan A (Amorim et al. 2016), polysaccharide (Chen et al. 2008), porphyran (Venkatpurwar et al. 2011), fucoxanthin (Singh et al. 2015), laminarin (Vijayakumar et al. 2022), sodium alginate (Sangeetha et al. 2012), gemcitabine (González-Ballesteros et al. 2023), etc.) or extracting, purifying, and subjecting pure compounds to synthesize nanoparticles could help researchers to understand the effective reducing, stabilizing, and caping agents with specific chemical functional groups present in the marine plants. Thereby, more detailed information can be generated on the enhanced bioactive properties of several drugs when combined with nanoparticles.

1.11 Limitations of Using Marine Plants for Synthesis of Nanoparticles

Unlike microbial resources (Bahrulolum et al. 2021) and terrestrial plants (Poudel et al. 2022), the utilization of marine plants for green synthesis of nanoparticles is limited by various factors such as abundance and accessibility of marine plants. For instance, unlike mangroves, certain seaweeds and seagrasses are not abundantly available throughout the year for collection in different geographical regions. Also, collection of interested seaweed species is not possible as seaweeds grow seasonally. Also, for another instance, the red seaweed, *Asparagopsis taxiformis* mostly found on reef environments is not available abundantly in intertidal regions. The only option is to collect the large amount of interested seaweed species during their

1.13 Conclusion

growth in a specific season and store them in −80 °C until further analyses. In some cases, seagrass species collection involves underwater diving, which is costly and time consuming to collect samples. Similarly, mangrove *Xylocarpus granatum* is not found in all the geographical locations. Some species of seaweeds (Venil et al. 2022) and seagrasses (Ramesh et al. 2020), are washed ashore in blooms/large quantities due to eutrophication and wind waves, respectively. In such instances, those marine plant resources can be collected, stored, and utilized for a wide range of experimental studies, including synthesizing nanoparticles.

1.12 Research Gaps

Green synthesis of a variety of metal and metal oxide nanoparticles has been successful and proven to show promising biological properties against various pathogenic microorganisms and tumor cells. Studies on marine plant derived nanoparticles conjugated with other bioactive compounds (phytocompounds/marine drugs) has been least tested or scarce. Thus, targeted research on the identification of key component or molecule with reducing, capping, and stabilizing properties has to be evaluated for better research as well as to avoid time to revalidation of the experiments with generalized conclusions (e.g., plant derived or genus derived or crude extract derived NPs). While, the optimized factors that impact the size and shape of nanoparticles derived from marine plants are least studied, indicating further research to be expanded in this direction. The biocompatibility of synthesized nanoparticles in many studies have not been recorded by hemolysis test, leavening a research gap on the negative impacts of known nanoparticles on environment and health. Randomly testing nanoparticle volumes (many researchers have used non-standard testing concentrations, deviating from established benchmarks like MIC, IC_{50}, LD_{50}, etc.) against test organisms hinders comparison and evaluation of their potential as nanodrugs, hindering meaningful comparisons across studies. Thus, adhering to standardized test concentrations allows for direct comparison and validation of novel results with established literature on the same pathogens, cancer cells, or model organisms. Finally, the toxicity of synthesized nanoparticles utilized in various fields and released into aquatic ecosystems (Banu et al. 2021), needs to be evaluated *in vitro* to understand the ecotoxicity of nanoparticle remnants on the aquatic biota at nanoscale.

1.13 Conclusion

Marine plants have long been recognised for their medicinal value, exhibiting a wide range of biological properties. The green-synthesized nanoparticles derived from such important medicinal marine plants have shown potential as bioactive agents for drug delivery, particularly in combating numerous multidrug-resistant microbial infections and tumor progression. Due to their inherent stability and safety, marine plants derived nanoparticles are promising candidates for further drug development aimed at targeted delivery to treat various infections. While, red seaweeds and mangroves are identified as promising sources of marine plants for nano drugs. Furthermore, silver nanoparticles are identified as promising drugs to treat aquaculture pathogens, including both bacterial and viral agents. However, the full therapeutic potential of nanoparticles derived from various marine plants remains underexplored, primarily due to limitations in research facilities and ethical considerations. Addressing these gaps through continued research is essential to unlock their industrial and biomedical applications.

References

Y. Abboud, T. Saffaj, A. Chagraoui, A.E. Bouari, K. Brouzi, O. Tanane, B. Ihssane, Biosynthesis, characterization and antimicrobial activity of copper oxide nanoparticles (CONPs) produced using brown alga extract (*Bifurcaria bifurcata*). Appl. Nanosci. **4**, 571–576 (2014)

N. Abdel-Raouf, N.M. Al-Enazi, I.B.M. Ibraheem, Green biosynthesis of gold nanoparticles using *Galaxaura elongata* and characterization of their antibacterial activity. Arab. J. Chem. **10**, S3029–S3039 (2017)

N. Abdel-Raouf, N.M. Al-Enazi, I.B.M. Ibraheem, R.M. Alharbi, M.M. Alkhulaifi, Biosynthesis of silver nanoparticles by using of the marine brown alga *Padina pavonia* and their characterization. Saudi J. Biol. Sci. **26**, 1207–1215 (2019)

N. Abdel-Raouf, R.M. Alharbi, N.M. Al-Enazi, M.M. Alkhulaifi, B.M.I. Ibraheem, Rapid biosynthesis of silver nanoparticles using the marine red alga *Laurencia catarinensis* and their characterization. Beni-Suef University J. Basic Appl. Sci. **7**, 150–157 (2018)

V. Abdi, I. Sourinejad, M. Yousefzadi, Z. Ghasemi, Biosynthesis of silver nanoparticles from the mangrove *Rhizophora mucronata*: Its characterization and antibacterial potential. Iran J. Sci. Technol. Trans. Sci. **43**, 2163–2171 (2019a)

V. Abdi, I. Sourinejad, M. Yousefzadi, Z. Ghasemi, Biosynthesis of Silver Nanoparticles from the Mangrove *Rhizophora mucronata*: its characterization and antibacterial potential. Iranian J. Sci. Technology. Trans. A: Sci. **43**, 2163–2171 (2019b)

V. Abdi, I. Sourinejad, M. Yousefzadi, Z. Ghasemi, Mangrove-mediated synthesis of silver nanoparticles using native *Avicennia marina* plant extract from southern Iran. Chem. Eng. Commun. **205**, 1069–1076 (2018)

N.M. Aboeita, S.A. Fahmy, M.M.H. El-Sayed, H.M.E.-S. Azzazy, T. Shoeib, Enhanced anticancer activity of nedaplatin loaded onto copper nanoparticles synthesized using red algae. Pharmaceutics **14**, 418 (2022)

E.F. Aboelfetoh, R.A. El-Shenody, M.M. Ghobara, Eco- friendly synthesis of silver nanoparticles using green algae (*Caulerpa serrulata*): reaction optimization, catalytic and antibacterial activities. Environ. Monit. Assess. **189**, 349 (2017)

S.E. Abo-Neima, A.A. Ahmed, M. El-Sheekh, M.E.M. Makhlof, *Polycladia myrica*-based delivery of selenium nanoparticles in combination with radiotherapy induces potent in vitro antiviral and in vivo anticancer activities against Ehrlich ascites tumor. Front. Mol. Biosci. **10**, 1120422 (2023)

D. Acharya, S. Satapathy, K.K. Yadav, P. Somu, G. Mishra, Systemic evaluation of mechanism of cytotoxicity in human colon cancer HCT-116 cells of silver nanoparticles synthesized using marine algae *Ulva lactuca* Extract. J. Inorg. Organomet. Polym. Mater. **32**, 596–605 (2022)

References

N.K. Ahila, V.S. Ramkumar, S. Prakash, B. Manikandan, J. Ravindran, P.K. Dhanalakshmi, E. Kannapiran, Synthesis of stable nanosilver particles (AgNPs) by the proteins of seagrass *Syringodium isoetifolium* and its biomedicinal properties. Biomed. Pharmacother. **84**, 60–70 (2016)

S. Ahmad, S. Munir, N. Zeb, A. Ullah, B. Khan, J. Ali, M. Bilal, M. Omer, M. Alamzeb, S.M. Salman, S. Ali, Green nanotechnology: a review on green synthesis of silver nanoparticles—an ecofriendly approach. Int. J. Nanomed. **14**, 5087–5107 (2019)

J.S. Ajarem, S.N. Maodaa, A.A. Allam, M.M. Taher, M. Khalaf, Benign synthesis of cobalt oxide nanoparticles containing red algae extract: antioxidant, antimicrobial, anticancer, and anticoagulant activity. J. Clust. Sci. **33**, 717–728 (2022)

Z. Ajdari, H. Rahman, K. Shameli, R. Abdullah, M.A. Ghani, S. Yeap, S. Abbasiliasi, D. Ajdari, A. Ariff, Novel Gold Nanoparticles Reduced by *Sargassum glaucescens*: Preparation. Characterization Anticancer Activity Molecules **21**, 123 (2016)

W.M. Alarif, Y.A. Shaban, M.I. Orif, M.A. Ghandourah, A.J. Turki, H.S. Alorfi, H.R.Z. Tadros, Green synthesis of TiO_2 nanoparticles using natural marine extracts for antifouling activity. Mar. Drugs **21**, 62 (2023)

R. Algotiml, A. Gab-alla, R. Seoudi, H.H. Abulreesh, I. Ahmad, K. Elbanna, Anticancer and antimicrobial activity of Red Sea seaweeds extracts-mediated gold nanoparticles. J. Pure Appl. Microbiol. **16**, 207–225 (2022a)

R. Algotiml, A. Gab-Alla, R. Seoudi, H.H. Abulreesh, M.Z. El-Readi, K. Elbanna, Anticancer and antimicrobial activity of biosynthesized Red Sea marine algal silver nanoparticles. Sci. Rep. **12**, 2421 (2022b)

S.M. Ali, V. Anuradha, N. Yogananth, R. Rajathilagam, A. Chanthuru, S.M. Marzook, Green synthesis of Silver nanoparticle by *Acanthus ilicifolius* mangrove plant against *Armigeres subalbatus* and *Aedes aegypti* mosquito larvae. Int. J. Nano Dimension **6**, 197–204 (2015)

A.L. Al-Malki, In vitro cytotoxicity and pro-apoptotic activity of phycocyanin nanoparticles from *Ulva lactuca* (Chlorophyta) algae. Saudi J. Biol. Sci. **27**, 894–898 (2020)

S. AlNadhari, N.M. Al-Enazi, F. Alshehrei, F. Ameen, A review on biogenic synthesis of metal nanoparticles using marine algae and its applications. Environ. Res. **194**, 110672 (2021)

M.S. Alsaggaf, A.M. Diab, B.E.F. ElSaied, A.A. Tayel, S.H. Moussa, Application of ZnO nanoparticles phycosynthesized with *Ulva fasciata* extract for preserving peeled shrimp quality. Nanomaterials **11**, 385 (2021)

S.A. Alsareii, A.M. Alamri, M.Y. AlAsmari, M.A. Bawahab, M.H. Mahnashi, I.A. Shaikh, A.K. Shettar, J.H. Hoskeri, V. Kumbar, Synthesis and characterization of silver nanoparticles from *Rhizophora apiculata* and studies on their wound healing, antioxidant, anti-inflammatory, and cytotoxic activity. Molecules **27**, 6306 (2022)

F.A. Alshubaily, E.J. Jambi, S.M. Khojah, M.J. Balgoon, M.H. Al-Zahrani, N.A. Alkhattabi, Studying the effect of silver nanoparticles synthesized by *Ulva fasciata* aqueous extract against liver toxicity induced by CCl4 in rats. J. Nature Sci. Med. **3**, 182–188 (2020)

M. Alvarez-Vinas, N. Gonzalez-Ballesteros, M.D. Torres, L. Lopez-Hortas, C. Vanini, G. Domingo, M.C. Rodríguez-Argüelles, H. Domínguez, Efficient extraction of carrageenans from *Chondrus crispus* for the green synthesis of gold nanoparticles and formulation of printable hydrogels. Int. J. Biol. Macromol. **206**, 553–566 (2022)

H.H. Amin, Biosynthesized silver nanoparticles using *Ulva lactuca* as a safe synthetic pesticide (in vitro). Open Agric **5**, 291–299 (2020)

M.O.R. Amorim, D.L. Gomes, L.A. Dantas, R.L.S. Viana, S.C. Chiquetti, J. Almeida-Lima, L.S. Costa, H.A.O. Rocha, Fucan-coated silver nanoparticles synthesized by a green method induce human renal adenocarcinoma cell death. Int. J. Biol. Macromol. **93**, 57–65 (2016)

V. Amutha, P. Deepak, C. Kamaraj, G. Balasubramani, D. Aiswarya, D. Arul, P. Santhanam, A.M. Ballamurugan, P. Perumal, Mosquito-larvicidal potential of metal and oxide nanoparticles synthesized from aqueous extract of the Seagrass, *Cymodocea serrulata*. J. Clust. Sci. **30**, 797–812 (2019)

R. Anjali, S. Palanisamy, M. Vinosha, A.M. Selvi, G. Sathiyaraj, T. Marudhupandi, S. Mohandoss, N.M. Prabhu, S. You, Fabrication of silver nanoparticles from marine macro algae *Caulerpa sertularioides*: characterization, antioxidant and antimicrobial activity. Process Biochem. **121**, 601–618 (2022)

J.J. Antony, P. Sivalingam, D. Siva, S. Kamalakkannan, K. Anbarasu, R. Sukirtha, M. Krishnan, S. Achiraman, Comparative evaluation of antibacterial activity of silver nanoparticles synthesized using *Rhizophora apiculata* and glucose. Colloids Surf. B Biointerfaces **88**, 134–140 (2011)

J.M.A. Antonysamy, S. Thangiah, R. Irulappan, Green synthesis of silver nanoparticles using *Dictyota bartayresiana* J.V. Lamouroux and their cytotoxic potentials. Int. Biol. Biomed. J. **1**, 112–118 (2015)

S. Anuluxan, A.C. Thavaranjit, S. Prabagar, R.C.L.D. Silva, J. Prabagar, Synthesis of silver nanoparticles from *Turbinaria ornata* and its antibacterial activity against water contaminating bacteria. Chem. Pap. **76**, 2365–2374 (2022)

D. Aravindan, T. Nallamuthu, R. Azeez, Green synthesis of silver nanoparticles using two seaweeds and their potential towards environment. Elixir Appl. Botany **76**, 28709–28715 (2014)

M.L.M.K. Arshan, S. Imaduddin, F. Magi, Biogenic synthesis of silver nanoparticles from mangrove plant *Lumnitzera racemosa* and its phytochemical screening & antibacterial activity. Asian J. Adv. Res. **3**, 64–71 (2020)

S. Arun, U. Saraswathi, Singaravelu, Green synthesis of silver nanoparticles using a mangrove *Excoecaria agallocha*. Int. J. Pharm. Sci. Invent. **3**, 54–57 (2014)

M. Arunkumar, K. Suhashini, N. Mahesh, R. Ravikumar, Quorum quenching and antibacterial activity of silver nanoparticles synthesized from *Sargassum polyphyllum*. Bangladesh J. Pharmacol. **9**, 54–59 (2014)

K.S. Asha, M. Johnson, K.P. Chandra, T. Shibila, I. Revathy, Extracellular synthesis of silver nanoparticles from a marine alga, *Sargassum polycystum* c. Agardh and their biopotentials. World J. Pharm. Pharm. Sci. **4**, 1388–1400 (2015)

R.M. Asik, B. Gowdhami, M.S.M. Jaabir, G. Archunan, N. Suganthy, Anticancer potential of zinc oxide nanoparticles against cervical carcinoma T cells synthesized via biogenic route using aqueous extract of *Gracilaria edulis*. Mater. Sci. Eng. C **103**, 109840 (2019)

N. Asmathunisha, K. Kathiresan, A review on biosynthesis of nanoparticles by marine organisms. Colloids Surf. B Biointerfaces **103**, 283–287 (2013)

M.N.A. Azeem, O.M. Ahmed, M. Shaban, K.N.M. Elsayed, In vitro antioxidant, anticancer, anti-inflammatory, anti-diabetic and anti-Alzheimer potentials of innovative macroalgae bio-capped silver nanoparticles. Environ. Sci. Pollut. Res. **29**, 59930–59947 (2022)

A.T. Aziz, Toxicity of *Ulva lactuca* and green fabricated silver nanoparticles against mosquito vectors and their impact on the genomic DNA of the dengue vector Aedes aegypti. IET Nanobiotechnol. **16**, 145–157 (2022)

S. Azizi, M.B. Ahmad, F. Namvar, R. Mohamad, Green biosynthesis and characterization of zinc oxide nanoparticles using brown marine macroalga *Sargassum muticum* aqueous extract. Mater. Lett. **116**, 275–277 (2014)

S. Azizi, F. Namvar, M. Mahdavi, M.B. Ahmad, R. Mohamad, Biosynthesis of silver nanoparticles using brown marine macroalga, *Sargassum muticum* aqueous extract. Materials (Basel) **6**, 5942–5950 (2013)

B. Babu, S. Palanisamy, M. Vinosha, R. Anjali, P. Kumar, B. Pandi, M. Tabarsa, S. You, N.M. Prabhu, Bioengineered gold nanoparticles from marine seaweed *Acanthophora spicifera* for pharmaceutical uses: antioxidant, antibacterial, and anticancer activities. Bioprocess. Biosyst. Eng. **43**, 2231–2242 (2020)

H. Bahrulolum, S. Nooraei, N. Javanshir, H. Tarrahimofrad, V.S. Mirbagheri, A.J. Easton, G. Ahmadian, Green synthesis of metal nanoparticles using microorganisms and their application in the agrifood sector. J. Nanobiotechnol. **19**, 86 (2021)

C. Baker, A. Pradhan, L. Pakstis, D.J. Pochan, S.I. Shah, Synthesis of antibacterial properties of silver nanoparticles. J. Nanosci. Nanotechnol. **5**, 244–249 (2005)

M. Bakshi, S. Ghosh, P. Chaudhuri, Green synthesis, characterization and antimicrobial potential of sliver nanoparticles using three mangrove plants from Indian Sundarban. Bionanoscience **5**, 162–170 (2015)

S. Balakrishnan, M. Srinivasan, J. Mohanraj, Biosynthesis of silver nanoparticles from mangrove plant (*Avicennia marina*) extract and their potential mosquito larvicidal property. J. Parasit. Dis. **40**, 991–996 (2016)

P. Balaraman, B. Balasubramanian, D. Kaliannan, M. Durai, H. Kamyab, S. Park, S. Chelliapan, C.T. Lee, V. Maluventhen, A. Maruthupandian, Phyco-synthesis of silver nanoparticles mediated from marine algae *Sargassum myriocystum* and its potential biological and environmental applications. Waste Biomass Valorization **11**, 5255–5271 (2020)

P. Balaraman, B. Balasubramanian, W.-C. Liu, D. Kaliannan, M. Durai, H. Kamyab, M. Alwetaishi, V. Maluventhan, V. Ashokkumar, S. Chelliapan, A. Maruthupandian, *Sargassum myriocystum* mediated TiO_2-nanoparticles and their antimicrobial, larvicidal activities and enhanced photocatalytic degradation of various dyes. Environ. Res. **204**, 112278 (2022)

A.N. Banu, N. Kudesia, A.M. Raut, I. Pakrudheen, J. Wahengbam, Toxicity, bioaccumulation, and transformation of silver nanoparticles in aqua biota: a review. Environ. Chem. Lett. **19**, 4275–4296 (2021)

A.A.B. Barnawi, S.E. Sharawi, J.A. Mahyoub, K.M. Al-Ghamdi, Larvicidal studies of *Avicennia marina* extracts against the dengue fever mosquito *Aedes aegypti* (Culicidae: Diptera). Int. J. Mosq. Res. **6**, 55–60 (2019)

G. Baskar, T. Palaniyandi, S. Viswanathan, M.R.A. Wahab, H. Surendran, M. Ravi, B.K. Rajendran, G. Govindasamy, A. Sivaji, S. Kaliamoorthy, Pharmacological effect of gold nanoparticles from red algae *Halymenia venusta* on A549 cell line. Inorg. Chem. Commun. **155**, 111005 (2023)

A.D.V. Bensy, G.J. Christobel, K. Muthusamy, A. Alfarhan, P. Anantharaman, Green synthesis of iron nanoparticles from *Ulva lactuca* and bactericidal activity against enteropathogens. J King Saud Univ Sci **34**, 101888 (2022)

B.V. Bhimba, J.S. Devi, Antimicrobial potential of silver nanoparticles synthesized using *Ulva reticulata*. Asian J. Pharm. Clin. Res. **7**, 82–85 (2014)

B.V. Bhimba, J.S. Devi, S.U. Nandhini, Green synthesis and cytotoxicity of silver nanoparticles from extracts of the marine macroalgae *Gracilaria corticata*. Indian J. Biotechnol. **14**, 276–281 (2015)

B.V. Bhimba, P.R. Kumari, Phytosynthesis of silver nanoparticles from the extracts of seaweed *Ulva lactuca* and its antimicrobial activity. Int. J. Pharm. Bio. Sci **5**, 666–677 (2014)

R. Bhuvaneswari, R.J. Xavier, M. Arumugam, Facile synthesis of multifunctional silver nanoparticles using mangrove plant *Excoecaria agallocha* L. for its antibacterial, antioxidant and cytotoxic effects. J. Parasit. Dis. **41**, 180–187 (2017)

P. Bhuyar, M.H.A. Rahim, S. Sundararaju, R. Ramaraj, G.P. Maniam, N. Govindan, Synthesis of silver nanoparticles using marine macroalgae *Padina* sp. and its antibacterial activity towards pathogenic bacteria. Beni Suef Univ. J. Basic Appl. Sci. **9**, 3 (2020)

B.E.E. Bialy, R.A. Hamouda, K.S. Khalifa, H.A. Hamza, Cytotoxic effect of biosynthesized silver nanoparticles on ehrlich ascites tumor cells in mice. Int. J. Pharmacol. **13**, 134–144 (2017)

L. Castro, M.L. Blázquez, J.A. Muñoz, F. González, A. Ballester, Biological synthesis of metallic nanoparticles using algae. IET Nanobiotechnol. 109–116 (2013). https://doi.org/10.1049/iet-nbt.2012.0041

N. Chakraborty, A. Banerjee, S. Lahiri, A. Panda, A.N. Ghosh, R. Pal, Biorecovery of gold using cyanobacteria and an eukaryotic alga with special reference to nanogold formation—a novel phenomenon. J. Appl. Phycol. **21**, 145–152 (2009)

A.B. Chanthini, G. Balasubramani, R. Ramkumar, R. Sowmiya, M.D. Balakumaran, P.T. Kalaichelvan, P. Perumal, Structural characterization, antioxidant and in vitro cytotoxic properties of seagrass, *Cymodocea serrulata* (R.Br.) Asch. & Magnus mediated silver nanoparticles. J. Photochem. Photobiol. B **153**, 145–152 (2015)

R. Chaudhary, K. Nawaz, A.K. Khan, C. Hano, B.H. Abbasi, S. Anjum, An overview of the algae-mediated biosynthesis of nanoparticles and their biomedical applications. Biomolecules **10**, 1498 (2020)

T. Chen, Y.-S. Wong, W. Zheng, Y. Bai, L. Huang, Selenium nanoparticles fabricated in *Undaria pinnatifida* polysaccharide solutions induce mitochondria-mediated apoptosis in A375 human melanoma cells. Colloids Surf. B Biointerfaces **67**, 26–31 (2008)

X. Chen, W. Han, X. Zhao, W. Tang, F. Wang, Epirubicin-loaded marine carrageenan oligosaccharide capped gold nanoparticle system for pH-triggered anticancer drug release. Sci. Rep. **9**, 6754 (2019)

X. Chen, X. Zhao, Y. Gao, J. Yin, M. Bai, F. Wang, Green synthesis of gold nanoparticles using Carrageenan Oligosaccharide and their in vitro antitumor activity. Mar. Drugs **16**, 277 (2018)

Z. Chengkui, C.K. Tseng, Z. Junfu, C.F. Chang, Chinese seaweeds in herbal medicine. Hydrobiologia **116**, 152–154 (1984)

S. Choudhary, V. Sangela, P. Saxena, V. Saharan, A. Pugazhendhi, Harish, Recent progress in algae-mediated silver nanoparticle synthesis. Int Nano Lett (2022)

D. Chugh, V.S. Viswamalya, B. Das, Green synthesis of silver nanoparticles with algae and the importance of capping agents in the process. J. Genetic Eng. Biotechnol. **19**, 126 (2021)

J.A. Colin, I.E. Pech-Pech, M. Oviedo, S.A. Águila, J.M. Romo-Herrera, O.E. Contreras, Gold nanoparticles synthesis assisted by marine algae extract: biomolecules shells from a green chemistry approach. Chem. Phys. Lett. **708**, 210–215 (2018)

L.H. Costa, J.V. Hemmer, E.H. Wanderlind, O.M.S. Gerlach, A.L.H. Santos, M.S. Tamanaha, A. Bella-Cruz, R. Corrêa, H.A.G. Bazani, C.M. Radetski, G.I. Almerindo, Green synthesis of gold nanoparticles obtained from algae *Sargassum cymosum*: optimization, characterization and stability. Bionanoscience **10**, 1049–1062 (2020)

S.A. Dahoumane, M. Mechouet, K. Wijesekera, C.D.M. Filipe, C. Sicard, D.A. Bazylinski, C. Jeffryes, Algae-mediated biosynthesis of inorganic nanomaterials as a promising route in nanobiotechnology—a review. Green Chem. **19**, 552 (2017)

S.K. Das, S. Behera, J.K. Patra, H. Thatoi, Green synthesis of sliver nanoparticles using *Avicennia officinalis* and *Xylocarpus granatum* extracts and in vitro evaluation of antioxidant, antidiabetic and anti-inflammatory activities. J. Clust. Sci. **30**, 1103–1113 (2019)

A.P. de Aragao, T.M. de Oliveira, P.V. Quelemes, M.L.G. Perfeito, M.C. Araujo, J.A.S. Santiago, V.S. Cardoso, P. Quaresma, J.R.S.A. Leite, D.A. da Silva, Green synthesis of silver nanoparticles using the seaweed *Gracilaria birdiae* and their antibacterial activity. Arabian J. Chem. **12**, 4182–4188 (2019)

P. Deepak, V. Amutha, R. Birundha, R. Sowmiya, C. Kamaraj, V. Balasubramanian, G. Balasubramani, D. Aiswarya, D. Arul, P. Perumal, Facile green synthesis of nanoparticles from brown seaweed *Sargassum wightii* and its biological application potential. Adv. Nat. Sci.: Nanosci. Nanotechnol. **9**, 035019 (2018)

P. Deepak, R. Sowmiya, R. Ramkumar, G. Balasubramani, D. Aiswarya, P. Perumal, Structural characterization and evaluation of mosquito-larvicidal property of silver nanoparticles synthesized from the seaweed, *Turbinaria ornata* (Turner) J. Agardh 1848. Artif. Cells Nanomed. Biotechnol. **45**, 990–998 (2017)

U.R. Devasena, R. Thiruchelvi, Synthesis of nanoparticles using algae and its application in dye degradation—a review. Int. J. Res. Anal. Rev. **6**, 252–256 (2019)

C.P. Devatha, A.K. Thalla, Green synthesis of nanomaterials, in *Synthesis of Inorganic Nanomaterials Advances and Key Technologies A Volume in Micro and Nano Technologies* ed. by S.M. Bhagyaraj, O.S. Oluwafemi, N. Kalarikkal, S. Thomas (Elsevier Ltd, 2018), pp. 169–184

J.S. Devi, B.V. Bhimba, Antibacterial and antifungal activity of silver nanoparticles synthesized using *Hypnea muciformis*. Biosci. Biotech. Res. Asia **11**, 235–238 (2014)

J.S. Devi, B.V. Bhimba, Anticancer activity of silver nanoparticles synthesized by the seaweed *Ulva lactuca* invitro. Open Access Sci. Rep. **1**, 1–5 (2012)

J.S. Devi, B.V. Bhimba, D.M. Peter, Production of biogenic Silver nanoparticles using *Sargassum longifolium* and its applications. Indian J. Mar. Sci. **42**, 125–130 (2013)

S.J. Devi, B.V. Bhimba, K. Ratnam, In vitro anticancer activity of silver nanoparticles synthesized using the extract of *Gelidiella* sp. Int. J. Pharm. Pharm. Sci. **4**, 710–715 (2012)

T.A. Devi, R.M. Sivaraman, S.S. Thavamani, T.P. Amaladhas, S. Devanesan, M.M. Kannan, Green synthesis of plasmonic nanoparticles using *Sargassum ilicifolium* and application in photocatalytic degradation of cationic dyes. Environ. Res. **208**, 112642 (2022)

P.K. Dhanalakshmi, R. Azeez, R. Rekha, S. Poonkodi, T. Nallamuthu, Synthesis of silver nanoparticles using green and brown seaweeds. Phykos **42**, 39–45 (2012)

S.P. Dhas, A. Mukherjee, N. Chandrasekaran, Phytosynthesis of silver nanoparticles using *Ceriops tagal* and its antimicrobial potential against human pathogens. Int. J. Pharm. Pharm. Sci. **5**, 349–352 (2013)

T.S. Dhas, V.G. Kumar, L.S. Abraham, V. Karthick, K. Govindaraju, *Sargassum myriocystum* mediated biosynthesis of gold nanoparticles. Spectrochim. Acta A Mol. Biomol. Spectrosc. **99**, 97–101 (2012)

T.S. Dhas, V.G. Kumar, V. Karthick, K.J. Angel, K. Govindaraju, Facile synthesis of silver chloride nanoparticles using marine alga and its antibacterial efficacy. Spectrochim. Acta A Mol. Biomol. Spectrosc. **120**, 416–420 (2014a)

T.S. Dhas, V.G. Kumar, V. Karthick, K. Govindaraju, T.S. Narayana, Biosynthesis of gold nanoparticles using *Sargassum swartzii* and its cytotoxicity effect on HeLa cells. Spectrochim. Acta A Mol. Biomol. Spectrosc. **133**, 102–106 (2014b)

T.S. Dhas, P. Sowmiya, K. Parthasarathy, A. Natarajan, G. Narendrakumar, R. Kumar, A.V. Samrot, S.U.M. Riyaz, V.K. Ganesh, V. Karthick, A. Rajasekar, In vitro antibacterial activity of biosynthesized silver nanoparticles against gram negative bacteria, in *Inorganic and Nano-Metal Chemistry* (2021)

D. Dixit, D. Gangadharan, K.M. Popat, C.R.K. Reddy, M. Trivedi, D.K. Gadhavi, Synthesis, characterization and application of green seaweed mediated silver nanoparticles (AgNPs) as antibacterial agents for water disinfection. Water Sci. Technol. Technol. **78**, 235–246 (2018)

T.N.J.I. Edison, R. Atchudan, C. Kamal, Y.R. Lee, *Caulerpa racemosa*: a marine green alga for eco-friendly synthesis of silver nanoparticles and its catalytic degradation of methylene blue. Bioprocess. Biosyst. Eng. **39**, 1401–1408 (2016)

H.S. El-Beltagi, A.A. Mohamed, H.I. Mohamed, K.M.A. Ramadan, A.A. Barqawi, A.T. Mansour, Phytochemical and potential properties of seaweeds and their recent applications: a review. Mar. Drugs **29**, 342 (2022)

H.Y. El-Kassas, M.A. Aly-Eldeen, S.M. Gharib, Green synthesis of iron oxide (Fe_3O_4) nanoparticles using two selected brown seaweeds: characterization and application for lead bioremediation. Acta Oceanol. Sin. **35**, 89–98 (2016)

H.Y. El-Kassas, M.M. ElKomi, Biogenic silver nanoparticles using seaweed *Ulva rigida* and their fungicidal and cytotoxic effects. JKAU Mar. Sci. **25**, 3–20 (2014)

H.Y. El-Kassas, M.M. El-Sheekh, Cytotoxic activity of biosynthesized gold nanoparticles with an extract of the red seaweed *Corallina officinalis* on the MCF-7 human breast cancer cell line. Asian Pac. J. Cancer Prev. **15**, 4311–4317 (2014)

H.Y. El-Kassas, M.G. Ghobrial, Biosynthesis of metal nanoparticles using three marine plant species: anti-algal efficiencies against "*Oscillatoria simplicissima*". Environ. Sci. Pollut. Res. **24**, 7837–7849 (2017)

A.Y. El-Khateeb, E.A. Hamed, F.Y. Ibrahim, S.E. Hamed, Eco-friendly synthesis of Selenium and Zinc nanoparticles with biocompatible *Sargassum latifolium* Algae extract in preservation of edible oils. J. Food Dairy Sci. Mansoura Univ. **10**, 141–146 (2019)

H.M. El-Rafie, M.H. El-Rafie, M.K. Zahran, Green synthesis of silver nanoparticles using polysaccharides extracted from marine macro algae. Carbohydr. Polym. **96**, 403–410 (2013)

A.A. Elrefaey, A.D. El-Gamal, S.M. Hamed, E.F. El-Belely, Algae-mediated biosynthesis of zinc oxide nanoparticles from *Cystoseira crinite* (Fucales; Sargassaceae) and it's antimicrobial and antioxidant activities. Egypt. J. Chem. **65**, 231–240 (2022)

M.M. El-Sheekh, H.Y. El-Kassas, Algal production of nano-silver and gold: their antimicrobial and cytotoxic activities: a review. J. Genetic Eng. Biotechnol. **14**, 299–310 (2016)

S.M. Etman, O.Y. Abdallah, Y.S.R. Elnaggar, Novel fucoidan based bioactive targeted nanoparticles from *Undaria pinnatifida* for treatment of pancreatic cancer. Int. J. Biol. Macromol. **145**, 390–401 (2020)

S.M. Etman, R.A. Mehanna, A.A. Bary, Y.S.R. Elnaggar, O.Y. Abdallah, *Undaria pinnatifida* fucoidan nanoparticles loaded with quinacrine attenuate growth and metastasis of pancreatic cancer. Int. J. Biol. Macromol. **170**, 284–297 (2021)

M. Faried, K. Shameli, M. Miyake, A. Hajalilou, K. Kalantari, Z. Zakaria, H. Hara, N.B.A. Khairudin, Synthesis of silver nanoparticles via green method using ultrasound irradiation in seaweed *Kappaphycus alvarezii* media. Res. Chem. Intermed. **42**, 7991–8004 (2016)

R. Fatima, M. Priya, L. Indurthi, V. Radhakrishnan, R. Sudhakaran, Biosynthesis of silver nanoparticles using red algae *Portieria hornemannii* and its antibacterial activity against fish pathogens. Microb. Pathog. **138**, 103780 (2020)

D. Fawcett, J.J. Verduin, M. Shah, S.B. Sharma, G.E.J. Poinern, A review of current research into the biogenic synthesis of metal and metal oxide nanoparticles via marine algae and seagrasses. J. Nanosci. **2017**, 8013850 (2017)

A. Fouda, A.M. Eid, A. Abdelkareem, H.A. Said, E.F. El-Belely, D.H.M. Alkhalifah, K.S. Alshallash, S.E.-D. Hassan, Phyco-synthesized Zinc Oxide nanoparticles using Marine Macroalgae, *Ulva fasciata* Delile, characterization, antibacterial activity, photocatalysis, and tanning wastewater treatment. Catalysts **12**, 756 (2022a)

A. Fouda, A.M. Eid, M.A. Abdel-Rahman, E.F. EL-Belely, M.A. Awad, S.E.D. Hassan, Z.E. AL-Faifi, M.F. Hamza, Enhanced antimicrobial, cytotoxicity, larvicidal, and repellence activities of Brown Algae, *Cystoseira crinita*-mediated green synthesis of magnesium oxide nanoparticles. Front Bioeng. Biotechnol. **10**, 849921 (2022b)

M. Francavilla, A. Pineda, A.A. Romero, J.C. Colmenares, C. Vargas, M. Monteleonea, R. Luque, Efficient and simple reactive milling preparation of photocatalytically active porous ZnO nanostructures using biomass derived polysaccharides. Green Chem. **16**, 2876 (2014)

V. Ganesan, D.J. Aruna, A. Astalakshmi, P. Nima, A. Thangaraja, Eco-friendly synthesis of silver nanoparticles using a sea weed, *Kappaphycus alvarezii* (Doty) Doty ex P.C.Silva. Int. J. Eng. Adv. Technol. **2**, 559–563 (2013)

M. Ghaemi, S. Gholamipour, Controllable synthesis and characterization of silver nanoparticles using *Sargassum angostifolium*. Iran. J. Chem. Chem. Eng. **36**, 1–10 (2017)

G. Ghodake, D.S. Lee, Biological synthesis of gold nanoparticles using the aqueous extract of the brown algae *Laminaria Japonica*. J. Nanoelectron. Optoelectron. **6**, 268–271 (2011)

M. Gnanadesigan, M. Anand, S. Ravikumar, M. Maruthupandy, M.S. Ali, V. Vijayakumar, A.K. Kumaraguru, Antibacterial potential of biosynthesised silver nanoparticles using *Avicennia marina* mangrove plant. Appl. Nanosci. **2**, 143–147 (2012)

M. Gnanadesigan, M. Anand, S. Ravikumar, M. Maruthupandy, V. Vijayakumar, S. Selvam, M. Dhineshkumar, A.K. Kumaraguru, Biosynthesis of silver nanoparticles by using mangrove plant extract and their potential mosquito larvicidal property. Asian Pac. J. Trop. Med. **4**, 799–803 (2011)

N. González-Ballesteros, L. Diego-González, M. Lastra-Valdor, M. Grimaldi, A. Cavazza, F. Bigi, M.C. Rodríguez-Argüelles, R. Simón-Vázquez, Immunomodulatory and antitumoral activity of gold nanoparticles synthesized by red algae aqueous extracts. Mar. Drugs **20**, 182 (2022)

N. González-Ballesteros, L. Diego-Gonzalez, M. Lastra-Valdor, M.C. Rodrıguez-Arguelles, M. Grimaldi, A. Cavazza, F. Bigi, R. Simon-Vazquez, Immunostimulant and biocompatible gold and silver nanoparticles synthesized using the *Ulva intestinalis* L. aqueous extract. J. Mater. Chem. B **7**, 4677 (2019)

N. González-Ballesteros, J.B. González-Rodríguez, M.C. Rodríguez-Argüelles, M. Lastra, New application of two Antarctic macroalgae *Palmaria decipiens* and *Desmarestia menziesii* in the synthesis of gold and silver nanoparticles. Polar Sci. **15**, 49–54 (2018)

N. González-Ballesteros, I. Maietta, R. Rey-Méndez, M.C. Rodríguez-Argüelles, M. Lastra-Valdor, A. Cavazza, M. Grimaldi, F. Bigi, R. Simón-Vázquez, Gold nanoparticles synthesized by an aqueous extract of *Codium tomentosum* as potential antitumoral enhancers of gemcitabine. Mar. Drugs **21**, 20 (2023)

N. González-Ballesteros, S. Prado-López, J.B. Rodríguez-González, M. Lastra, M.C. Rodríguez-Argüelles, Green synthesis of gold nanoparticles using brown algae *Cystoseira baccata*: Its activity in colon cancer cells. Colloids Surf. B Biointerfaces **153**, 190–198 (2017)

N. González-Ballesteros, M.C. Rodríguez-Argüelles, M. Grimaldi, A. Cavazza, F. Bigi, Synthesis of silver and gold nanoparticles by *Sargassum muticum* biomolecules and evaluation of their antioxidant activity and antibacterial properties. J Nanostructure Chem **10**, 317–330 (2020)

N. González-Ballesteros, M.C. Rodríguez-Argüelles, M. Lastra-Valdor, Evaluation of the antioxidant capacities of antarctic macroalgae and their use for nanoparticles production. Molecules **26**, 1182 (2021)

N. González-Ballesteros, M.C. Rodríguez-Argüelles, S. Prado-López, M. Lastra, M. Grimaldi, A. Cavazza, L. Nasi, G. Salviati, F. Bigi, Macroalgae to nanoparticles: study of *Ulva lactuca* L. role in biosynthesis of T gold and silver nanoparticles and of their cytotoxicity on colon cancer cell lines. Mater. Sci. Eng. C **97**, 498–509 (2019b)

M. Gopu, P. Kumar, T. Selvankumar, B. Senthilkumar, C. Sudhakar, M. Govarthanan, R.S. Kumar, K. Selvam, Green biomimetic silver nanoparticles utilizing the red algae *Amphiroa rigida* and its potent antibacterial, cytotoxicity and larvicidal efficiency. Bioprocess. Biosyst. Eng. **44**, 217–223 (2021)

S. Gouda, R.G. Kerry, G. Das, J.K. Patra, Synthesis of nanoparticles utilizing sources from the Mangrove environment and their potential applications—an overview, in *Nanomaterials in Plants, Algae, and Microorganisms* (Elsevier, 2019), pp. 219–235

K. Govindaraju, V. Kiruthiga, V.G. Kumar, G. Singaravelu, Extracellular synthesis of silver nanoparticles by a marine alga, *Sargassum wightii* Grevilli and their antibacterial effects. J. Nanosci. Nanotechnol. **9**, 5497–5501 (2009)

K. Govindaraju, K. Krishnamoorthy, S.A. Alsagaby, G. Singaravelu, M. Premanathan, Green synthesis of silver nanoparticles for selective toxicity towards cancer cells. IET Nanobiotechnol. **9**, 325–330 (2015)

H. Gu, X. Chen, F. Chen, X. Zhou, Z. Parsaee, Ultrasound-assisted biosynthesis of CuO-NPs using brown alga *Cystoseira trinodis*: characterization, photocatalytic AOP, DPPH scavenging and antibacterial investigations. Ultrason. Sonochem. **41**, 109–119 (2018)

S. Gurusamy, M.R. Kulanthaisamy, D.G. Hari, A. Veleeswaran, B. Thulasinathan, J.B. Muthuramalingam, R. Balasubramani, S.W. Chang, M.V. Arasu, N.A. Al-Dhabi, A. Selvaraj, A. Alagarsamy, Environmental friendly synthesis of TiO_2-ZnO nanocomposite catalyst and silver nanomaterials for the enhanced production of biodiesel from *Ulva lactuca* seaweed and potential antimicrobial properties against the microbial pathogens. J. Photochem. Photobiol. B **193**, 118–130 (2019)

R.A. Hamouda, M.A. El Mongy, K.F. Eid, Comparative study between two red algae for biosynthesis silver nanoparticles capping by SDS: insights of characterization and antibacterial activity. Microb. Pathog. **129**, 224–232 (2019)

R.A.E.F. Hamouda, M.A. El-Mongy, K.F. Eid, Antibacterial activity of Silver nanoparticles using *Ulva fasciata* extracts as reducing agent and sodium dodecyl sulfate as stabilizer. Int. J. Pharmacol. **14**, 359–368 (2018)

S. Hashemi, M.H. Givianrad, A.M. Moradi, K. Larijani, Biosynthesis of silver nanoparticles using brown marine seaweed *Padina boeregeseni* and evaluation of physico-chemical factors. Indian J. Geo-Marine Sci. **44**, 1415–1421 (2015)

I. Hussain, N.B. Singh, A. Singh, H. Singh, S.C. Singh, Green synthesis of nanoparticles and its potential application. Biotechnol. Lett. **38**, 545–560 (2016)

R.R.S. Hussein, A.A. Farghali, A.H.A. Hassanein, I.B.M. Ibraheem, Biosynthesis of silver nanoparticles by using of the marine alga *Gracilaria parvispora* and its antagonistic efficacy against some common skin infecting pathogens. Aust. J. Basic Appl. Sci. **11**, 219–227 (2017)

I.B.M. Ibraheem, E.B.E.E. Abd, W.F. Saad, W.A. Fathy, Green biosynthesis of silver nanoparticles using marine red algae *Acanthophora specifera* and its antimicrobial activity. J. Nanomed. Nanotechnol. **7**, 1000409 (2016)

G. Isaac, R.E. Renitta, Brown algae mediated synthesis, characterization of gold nano particles using *Padina pavonica* and their antibacterial activity against human pathogens. Int. J. Pharmtech. Res. **8**, 31–40 (2015)

R. Ishwarya, B. Vaseeharan, S. Kalyani, B. Banumathi, M. Govindarajan, N.S. Alharbi, S. Kadaikunnan, M.N. Al-anbr, J.M. Khaled, G. Benelli, Facile green synthesis of zinc oxide nanoparticles using *Ulva lactuca* seaweed extract and evaluation of their photocatalytic, antibiofilm and insecticidal activity. J. Photochem. Photobiol. B **178**, 249–258 (2018a)

R. Ishwarya, B. Vaseeharan, S. Subbaiah, A.K. Nazar, M. Govindarajan, N.S. Alharbi, S. Kadaikunnan, J.M. Khaled, M.N. Al-anbr, *Sargassum wightii*-synthesized ZnO nanoparticles—from antibacterial and insecticidal activity to immunostimulatory effects on the green tiger shrimp *Penaeus semisulcatus*. J. Photochem. Photobiol. B **183**, 318–330 (2018b)

A.M. Ismail, H.S.J. Ali, M. Parthasarathy, Biosynthesis of gold nano-particles using *Sargassum myriocystum* and evaluation of their antibacterial activity. Int. J. Pure App. Biosci. **6**, 1340–1350 (2018)

R.H. Jacob, S.M. Shanab, E.A. Shalaby, Algal biomass nanoparticles: chemical characteristics, biological actions, and applications. Biomass. Convers. Biorefin. (2021). https://doi.org/10.1007/s13399-021-01930-y

S. Jadoun, R. Arif, N.K. Jangid, R.K. Meena, Green synthesis of nanoparticles using plant extracts: a review. Environ. Chem. Lett. **19**, 355–374 (2020)

P. Jegadeeswaran, R. Shivaraj, R. Venckatesh, Green synthesis of silver nanoparticles from extract of *Padina tetrastromatica* leaf. Dig. J. Nanomater. Biostruct. **7**, 991–998 (2012)

G.-J. Jeong, S. Khan, N. Tabassum, F. Khan, Y.-M. Kim, Marine-bioinspired nanoparticles as potential drugs for multiple biological roles. Mar. Drugs **20**, 527 (2022)

S. Jeyarani, N.M. Vinita, P. Puja, S. Senthamilselvi, U. Devan, A.J. Velangani, M. Biruntha, A. Pugazhendhi, P. Kumar, Biomimetic gold nanoparticles for its cytotoxicity and biocompatibility evidenced by fluorescence-based assays in cancer (MDA-MB-231) and non-cancerous (HEK-293) cells. J. Photochem. Photobiol. b. **202**, 111715 (2020)

E.-S. Jun, Y.J. Kim, H.-H. Kim, S.Y. Park, Gold nanoparticles using *Ecklonia stolonifera* protect human dermal fibroblasts from UVA-induced senescence through inhibiting MMP-1 and MMP-3. Mar. Drugs **18**, 433 (2020)

S. Kaidi, Z. Belattmania, F. Bentiss, C. Jama, A. Reani, B. Sabour, Synthesis and characterization of silver nanoparticles using alginate from the brown seaweed *Laminaria ochroleuca*: structural features and antibacterial activity. Biointerface Res. Appl. Chem. **12** (2022)

S. Kailasam, A. Sundaramanickam, R. Tamilvanan, S.V. Kanth, Macrophytic waste optimization by synthesis of silver nanoparticles and exploring their agro- fungicidal activity. Inorganic Nano-Metal Chem. **53**, 257–266 (2023)

K. Kalimuthu, C. Panneerselvam, C. Chou, S.-M. Lin, L.-C. Tseng, K.-H. Tsai, K. Murugan, J.-S. Hwang, Predatory efficiency of the copepod *Megacyclops formosanus* and toxic effect of the red alga *Gracilaria firma*-synthesized silver nanoparticles against the dengue vector *Aedes aegypti*. Hydrobiologia **785**, 359–372 (2017)

M. Kamal, N. Abdel-Raouf, H. Sonbol, H. Abdel-Tawab, M.S. Abdelhameed, O. Hammouda, K.N.M. Elsayed, In vitro assessment of antimicrobial, anti-inflammatory, and schistolarvicidal activity of macroalgae-based gold nanoparticles. Front. Mar. Sci. **9**, 1075832 (2022)

C. Kamaraj, S. Karthi, A.D. Reegan, G. Balasubramani, G. Ramkumar, K. Kalaivani, A.A. Zahir, P. Deepak, S. Senthil-Nathan, M.M. Rahman, A.R.M.T. Islam, G. Malafaia, Green synthesis of gold nanoparticles using *Gracilaria crassa* leaf extract and their ecotoxicological potential: issues to be considered. Environ. Res. **213**, 113711 (2022)

R. Kannan, Ragupathi Raja, R. Arumugam, D. Ramya, K. Manivannan, P. Anantharaman, Green synthesis of silver nanoparticles using marine macroalga *Chaetomorpha linum*. Appl. Nanosci. **3**, 229–233 (2013)

R.R.R. Kannan, W.A. Stirk, J.V. Staden, Synthesis of silver nanoparticles using the seaweed *Codium capitatum* P.C. Silva (Chlorophyceae). S. Afr. J. Bot. **86**, 1–4 (2013b)

H.Y.E. Kassas, A.A. Attia, Bactericidal application and cytotoxic activity of biosynthesized silver nanoparticles with an extract of the red seaweed *Pterocladiella capillacea* on the HepG2 cell line. Asian Pac. J. Cancer Prev. **15**, 1299–1306 (2014)

T. Kathiraven, A. Sundaramanickam, N. Shanmugam, T. Balasubramanian, Green synthesis of silver nanoparticles using marine algae *Caulerpa racemosa* and their antibacterial activity against some human pathogens. Appl. Nanosci. **5**, 499–504 (2015)

K. Kathiresan, N.M. Alikunhi, A. Nabikhan, In vitro synthesis of antimicrobial silver nanoparticles by mangroves, saltmarshes and plants of coastal origin. Int. J. Biomedical Nanoscience and Nanotechnology **2**, 284–298 (2012)

K. Kayalvizhi, N. Asmathunisha, V. Subramanian, K. Kathiresan, Purification of silver and gold nanoparticles from two species of brown seaweeds (*Padina tetrastromatica* and *Turbinaria ornata*). J. Med. Plants Stud. **2**, 32–37 (2014)

M.M.S.I. Khaleelullah, M. Murugan, K.V. Radha, D. Thiyagarajan, Y. Shimura, Y. Hayakawa, Synthesis of super-paramagnetic iron oxide nanoparticles assisted by brown seaweed *Turbinaria decurrens* for removal of reactive navy blue dye. Mater. Res. Express **4**, 105038 (2017)

K.S. Khalifa, R.A. Hamouda, D.H.A. Hamza, In vitro antitumor activity of silver nanoparticles biosynthesized by marine algae. Dig. J. Nanomater. Biostruct. **11**, 213–221 (2016)

H.S. Khalilieh, A. Boulos, A glimpse on the uses of seaweeds in Islamic science and daily life during the classical period. Arab. Sci. Philos. **16**, 91–101 (2006)

F. Khan, P. Manivasagan, J.-W. Lee, D.T.N. Pham, J. Oh, Y.-M. Kim, Fucoidan-stabilized gold nanoparticle-mediated biofilm inhibition, attenuation of virulence and motility properties in *Pseudomonas aeruginosa* PAO1. Mar. Drugs **17**, 208 (2019)

K.D. Khan, U. Hanif, I. Liaqat, S. Shaheen, U.F. Awan, S. Ishtiaq, L. Pereira, S. Bahadur, M.D. Khan, Application of green silver nanoparticles synthesized from the red seaweeds *Halymenia porphyriformis* and *Solieria robusta* against oral pathogenic bacteria by using microscopic technique. Front Biosci. **14**, 13 (2022a)

M.S. Khan, P.P. Dhavan, B.L. Jadhav, N.G. Shimpi, Ultrasound-assisted green synthesis of Ag-decorated ZnO nanoparticles using *Excoecaria agallocha* leaf extract and evaluation of their photocatalytic and biological activity. Energy Technol. Environ. Sci. **5**, 12660–12671 (2020)

M.S. Khan, S. Ranjani, S. Hemalatha, Synthesis and characterization of *Kappaphycus alvarezii* derived silver nanoparticles and determination of antibacterial activity. Mater. Chem. Phys. **282**, 125985 (2022b)

H. Khanehzaei, M.B. Ahmad, K. Shameli, Z. Ajdari, Synthesis and characterization of Cu@Cu2O core shell nanoparticles prepared in Seaweed *Kappaphycus alvarezii* Media. Int. J. Electrochem. Sci. **10**, 404–413 (2015)

P. Khanna, A. Kaur, D. Goyal, Algae-based metallic nanoparticles: synthesis, characterization and applications. J. Microbiol. Methods **163**, 105656 (2019)

D.-Y. Kim, R.G. Saratale, S. Shinde, A. Syed, F. Ameen, G. Ghodake, Green synthesis of silver nanoparticles using *Laminaria japonica* extract: characterization and seedling growth assessment. J. Clean. Prod. **172**, 2910–2918 (2018)

A. Kingslin, K. Kalimuthu, M.L. Kiruthika, A.S. Khalifa, P.T. Nhat, K. Brindhadevi, Synthesis, characterization and biological potential of silver nanoparticles using Enteromorpha prolifera algal extract. Appl Nanosci. (2022a)

A. Kingslin, K. Kalimuthu, P. Viswanathan, Nanocatalytic efficacy of silver nanoparticles fabricated using *Chaetomorpha antennina* algal extract, their characterization, and its applications. J. Sci. Res. **14**, 343–362 (2022b)

A. Kingslin, P. Ravikumar, Green synthesis, characterization and applications of silver nanoparticles of *Padina tetrastromatica* Hauck. World J. Pharm. Pharm. Sci. **5**, 1304–1323 (2016)

M.V. Kiran, S. Murugesan, Biological synthesis of silver nanoparticles from marine alga *Colpomenia sinuosa* and its in vitro anti-diabetic activity. Am. J. Bio-Pharmacol. Biochem. Life Sci. **3**, 1–7 (2014a)

M.V. Kiran, S. Murugesan, Bio-synthesis of silver nano particles from marine alga *Halymenia poryphyroides* and its antibacterial efficacy. Int. J. Curr. Microbiol. App. Sci. **3**, 1–8 (2014b)

A.T. Koçer, D. Özçimen, Eco-friendly synthesis of silver nanoparticles from macroalgae: optimization, characterization and antimicrobial activity. Biomass Convers Biorefin (2022)

M.H.Z. Kochesfehani, S.A. Jaliseh, M.Z. Kochesfehani, Antibacterial effect of silver nanoparticles synthesized from the red algae *Gracilaria gracilis*. J. Microbial. World **13**, 369–378 (2021)

H. Koopi, F. Buazar, A novel one-pot biosynthesis of pure alpha aluminum oxide nanoparticles using the macroalgae *Sargassum ilicifolium*: a green marine approach. Ceram. Int. **44**, 8940–8945 (2018)

M. Krishnan, V. Sivanandham, D. Hans-Uwe, S.G. Murugaiah, P. Seeni, S. Gopalan, A.J. Rathinam, Antifouling assessments on biogenic nanoparticles: a field study from polluted offshore platform. Mar. Pollut. Bull. **101**, 816–825 (2015)

P. Kumar, M. Govindaraju, S. Senthamilselvi, K. Premkumar, Photocatalytic degradation of methyl orange dye using silver (Ag) nanoparticles synthesized from *Ulva lactuca*. Colloids Surf. B Biointerfaces **103**, 658–661 (2013a)

P. Kumar, S.S. Selvi, M. Govindaraju, Seaweed-mediated biosynthesis of silver nanoparticles using *Gracilaria corticata* for its antifungal activity against *Candida* spp. Appl. Nanosci. **3**, 495–500 (2013b)

P. Kumar, S.S. Selvi, A.L. Prabha, K.P. Kumar, R.S. Ganeshkumar, M. Govindaraju, Synthesis of silver nanoparticles from *Sargassum tenerrimum* and screening phytochemicals for its antibacterial activity. Nano Biomed. Eng. **4**, 12–16 (2012a)

P. Kumar, S.S. Selvi, A.L. Prabha, M. Selvaraj, L.M. Rani, P. Suganthi, B.S. Devi, M. Govindaraju, Antibacterial activity and in-vitro cytotoxicity assay against brine shrimp using silver nanoparticles synthesized from *Sargassum ilicifolium*. Dig. J. Nanomater. Biostruct. **7**, 1447–1455 (2012b)

P. Kumar, S. Senthamilselvi, A. Lakshmipraba, K. Premkumar, R. Muthukumaran, P. Visvanathan, R.S. Ganeshkumar, M. Govindaraju, Efficacy of bio-synthesized silver nanoparticles using *Acanthophora spicifera* to encumber biofilm formation. Dig. J. Nanomater. Biostruct. **7**, 511–522 (2012c)

P.S.M. Kumar, D. MubarakAli, R.G. Saratale, G.D. Saratale, A. Pugazhendhi, K. Gopalakrishnan, N. Thajuddin, Synthesis of nano-cuboidal gold particles for effective antimicrobial property against clinical human pathogens. Microb. Pathog. **113**, 68–73 (2017a)

S.D. Kumar, G. Singaravelu, S. Ajithkumar, K. Murugan, M. Nicoletti, G. Benelli, Mangrove-mediated green synthesis of silver nanoparticles with high HIV-1 reverse transcriptase inhibitory potential. J. Clust. Sci. **28**, 359–367 (2017b)

S.V. Kumar, S. Rajeshkumar, Optimized production of silver nanoparticles using marine macroalgae *Sargassum myriocystum* for its antibacterial activity. J. Bionanosci. **11**, 323–329 (2017)

V.A. Kumar, K. Ammani, R. Jobina, P. Parasuraman, B. Siddhardha, Larvicidal activity of green synthesized silver nanoparticles using *Excoecaria agallocha* L. (Euphorbiaceae) leaf extract against *Aedes aegypti*. IET Nanobiotechnol. **10**, 382–388 (2016)

M. Kumaresan, K.V. Anand, K. Govindaraju, S. Tamilselvan, V.G. Kumar, Seaweed *Sargassum wightii* mediated preparation of zirconia (ZrO_2) nanoparticles and their antibacterial activity against gram positive and gram negative bacteria. Microb. Pathog. **124**, 311–315 (2018)

V.S. Kumari, G.S. Sundari, S.K. Basha, Facile green synthesis of gold nanoparticles with great catalytic activity using *Ulva fasciata*. Lett. Appl. NanoBioScience **3**, 124–129 (2014)

H.E. Lashgarian, M. Karkhane, A.K. Alhameedawi, A. Marzban, Phyco-mediated synthesis of Ag/AgCl nanoparticles using ethanol extract of a marine green algae, *Ulva fasciata* Delile with biological activity. Biointerface Res. Appl. Chem. **11**, 14545–14554 (2021)

V. Lavakumar, K. Masilamani, V. Ravichandiran, N. Venkateshan, D.V.R. Saigopal, C.K.A. Kumar, C. Sowmya, Promising upshot of silver nanoparticles primed from *Gracilaria crassa* against bacterial pathogens. Chem. Cent. J. **9**, 42 (2015)

K.W. Lee, D. Jeong, K. Na, Doxorubicin loading fucoidan acetate nanoparticles for immune and chemotherapy in cancer treatment. Carbohydr. Polym. **94**, 850–856 (2013)

References

M.S. Lefteh, I. Sourinejad, Z. Ghasemi, Biosynthesis of Titanium Dioxide nanoparticles from the Mangrove (*Avicennia marina*) and investigation of its antibacterial activity. J. Mazandaran Univ. Med. Sci. **30**, 15–27 (2020)

S. Liang, F. Guo, S. Du, B. Tian, Y. Dong, X. Jia, L. Qian, Synthesis of *Sargassum* char-supported Ni-Fe nanoparticles and its application in tar cracking during biomass pyrolysis. Fuel **275**, 117923 (2020)

B. Liu, J. Xie, J.Y. Lee, Y.P. Ting, J.P. Chen, Optimization of high-yield biological synthesis of single-crystalline gold nanoplates. J. Phys. Chem. B **109**, 15256–15263 (2005)

K.-Y. Lu, R. Li, C.-H. Hsu, L. Cheng-Wei, S.-C. Chou, M.-L. Tsai, F.-L. Mi, Development of a new type of multifunctional fucoidan-based nanoparticles for anticancer drug delivery. Carbohydr. Polym. **165**, 410–420 (2017)

P. Madhiyazhagan, K. Murugan, A.N. Kumar, T. Nataraj, D. Dinesh, C. Panneerselvam, J. Subramaniam, P.M. Kumar, U. Suresh, M. Roni, M. Nicoletti, A.A. Alarfaj, A. Higuchi, M.A. Munusamy, G. Benelli, *Sargassum muticum*-synthesized silver nanoparticles: an effective control tool against mosquito vectors and bacterial pathogens. Parasitol. Res. **114**, 4305–4317 (2015)

P. Madhiyazhagan, K. Murugan, A.N. Kumar, T. Nataraj, J. Subramaniam, B. Chandramohan, C. Panneerselvam, D. Dinesh, U. Suresh, M. Nicoletti, M.S. Alsalhi, S. Devanesan, G. Benelli, One pot synthesis of silver nanocrystals using the seaweed *Gracilaria edulis*: biophysical characterization and potential against the filariasis vector *Culex quinquefasciatus* and the midge *Chironomus circumdatus*. J. Appl. Phycol. **29**, 649–659 (2017)

M. Mahdavi, F. Namvar, M.B. Ahmad, R. Mohamad, Green biosynthesis and characterization of magnetic iron oxide (Fe_3O_4) nanoparticles using seaweed (*Sargassum muticum*) aqueous extract. Molecules **18**, 5954–5964 (2013)

G.N. Maity, S. Mondal, Ag-nanoparticles based on polysaccharide isolated from the leaves of *Xylocarpus granatum* and their study on microbes and comparison with cipran 500. Asian J. Microbiol. Biotechnol. Environ. Exp. Sci. **19**, 397–403 (2017)

G.N. Maity, J. Sarkar, S. Khatua, S. Mondal, K. Acharya, Green synthesis of silver nanoparticles using mangrove fruit polysaccharide for bacterial growth inhibition. Asian J. Pharm. Clin. Res. **12**, 179–183 (2019)

V.K. Manam, M. Subbaiah, Biosynthesis and characterization of silver nanoparticles from marine macroscopic red seaweed *Halymenia porphyroides* boergesen (crypton) and its antifungal efficacy against dermatophytic and non-dermatophytic fungi. Asian J. Pharm. Clin. Res. **13**, 174–181 (2020)

R. Manikandan, R. Anjali, M. Beulaja, N.M. Prabhu, A. Koodalingam, G. Saiprasad, P. Chitra, M. Arumugam, Synthesis, characterization, anti-proliferative and wound healing activities of silver nanoparticles synthesized from *Caulerpa scalpelliformis*. Process Biochem. **79**, 135–141 (2019)

H.A. Mansouri-Tehrani, M. Keyhanfar, M. Behbahani, G. Dini, Synthesis and characterization of algae-coated selenium nanoparticles as a novel antibacterial agent against *Vibrio harveyi*, a *Penaeus vannamei* pathogen. Aquaculture **534**, 736260 (2021)

S. Mashjoor, M. Yousefzadi, H. Zolgharnain, E. Kamrani, M. Alishahi, Organic and inorganic nano-Fe_3O_4: alga *Ulva flexuosa*-based synthesis, antimicrobial effects and acute toxicity to briny water rotifer *Brachionus rotundiformis*. Environ. Pollut. **237**, 50–64 (2018)

A. Massironi, A. Morelli, L. Grassi, D. Puppi, S. Braccini, G. Maisetta, S. Esin, G. Batoni, C.D. Pina, F. Chiellini, Ulvan as novel reducing and stabilizing agent from renewable algal biomass: application to green synthesis of silver nanoparticles. Carbohydr. Polym. **203**, 310–321 (2019)

Y.N. Mata, E. Torres, M.L. Blázquez, A. Ballester, F. González, J.A. Munoz, Gold(III) biosorption and bioreduction with the brown alga *Fucus vesiculosus*. J. Hazard. Mater. **166**, 612–618 (2009)

X. Men, W. Xu, X. Zhu, W. Ma, Extraction, selenium-nanoparticle preparation and anti-virus bioactivity determination of polysaccharides from *Caulerpa taxifolia*. Zhong Yao Cai **32**, 1891–1894 (2009)

F.T. Minhas, G. Arslan, I.H. Gubbuk, C. Akkoz, B.Y. Ozturk, B. Asıkkutlu, U. Arslan, M. Ersoz, Evaluation of antibacterial properties on polysulfone composite membranes using synthesized biogenic silver nanoparticles with *Ulva compressa* (L.) Kutz. and *Cladophora glomerata* (L.) Kutz. extracts. Int. J. Biol. Macromol. **107**, 157–165 (2018)

M. Mmola, M.L. Roes-Hill, K. Durrell, J.J. Bolton, N. Sibuyi, M.E. Meyer, D.R. Beukes, E. Antunes, Enhanced antimicrobial and anticancer activity of silver and gold nanoparticles synthesised using *Sargassum incisifolium* aqueous extracts. Molecules **21**, 1633 (2016)

R.M. Mohamed, E.M. Fawzy, R.A. Shehab, M.O. Abdel-Salam, R.A.S.E. Din, H.M.A.E. Fatah, Production, characterization, and cytotoxicity effects of silver nanoparticles from brown alga (*Cystoseira myrica*). J. Nanotechnol. **2022**, 6469090 (2022)

C. Mohandass, A.S. Vijayaraj, R. Rajasabapathy, S. Satheeshbabu, S.V. Rao, C. Shiva, L. De-Mello, Biosynthesis of silver nanoparticles from marine seaweed *Sargassum cinereum* and their antibacterial activity. Indian J. Pharm. Sci. **75**, 606–610 (2013)

S. Momeni, I. Nabipour, A simple green synthesis of palladium nanoparticles with *Sargassum* alga and their electrocatalytic activities towards hydrogen peroxide. Appl. Biochem. Biotechnol. **176**, 1937–1949 (2015)

S. Momeni, A. Safavi, R. Ahmadi, I. Nabipour, Gold nanosheets synthesized with red marine alga *Actinotrichia fragilis* as efficient electrocatalysts toward formic acid oxidation. RSC Adv. **6**, 75152–75161 (2016)

M.S. Montasser, A.M. Younes, M.M. Hegazi, N.H. Dashti, A.E. El-Sharkawey, G.W. Beall, A novel eco-friendly method of using red algae (*Laurencia papillosa*) to synthesize gold nanoprisms. J Nanomed. Nanotechnol. **7**, 1000383 (2016)

P.V. Moorthi, C. Balasubramanian, S. Mohan, An improved insecticidal activity of silver nanoparticle synthesized by using *Sargassum muticum*. Appl. Biochem. Biotechnol. **175**, 135–140 (2015)

A. Moshfegh, A. Jalali, A. Salehzadeh, A.S. Jozani, Biological synthesis of silver nanoparticles by cell-free extract of *Polysiphonia* algae and their anticancer activity against breast cancer MCF-7 cell lines. Micro Nano Lett. **14**, 581–584 (2019)

O.C. Mukhoro, W.D. Roos, M. Jaffer, J.J. Bolton, M.J. Stillman, D.R. Beukes, E. Antunes, Very green photosynthesis of gold nanoparticles by a living aquatic plant: photoreduction of AuIII by the seaweed *Ulva armoricana*. Chem. Eur. J. **24**, 1657–1666 (2018)

V. Murugammal, G. Flora, Synthesis of silver nanopartices using red algae, characterization and effect on beneficial soil microbes. Asian J. Biol. Life Sci. **6**, 313–320 (2017)

K. Murugan, P. Aruna, C. Panneerselvam, P. Madhiyazhagan, M. Paulpandi, J. Subramaniam, R. Rajaganesh, H. Wei, M.S. Alsalhi, S. Devanesan, M. Nicoletti, B. Syuhei, A. Canale, G. Benelli, Fighting arboviral diseases: low toxicity on mammalian cells, dengue growth inhibition (in vitro), and mosquitocidal activity of *Centroceras clavulatum*-synthesized silver nanoparticles. Parasitol. Res. **115**, 651–662 (2016a)

K. Murugan, G. Benelli, S. Ayyappan, D. Dinesh, C. Panneerselvam, M. Nicoletti, J.-S. Hwang, P.M. Kumar, J. Subramaniam, U. Suresh, Toxicity of seaweed-synthesized silver nanoparticles against the filariasis vector *Culex quinquefasciatus* and its impact on predation efficiency of the cyclopoid crustacean *Mesocyclops longisetus*. Parasitol. Res. **114**, 2243–2253 (2015a)

K. Murugan, D. Dinesh, M. Paulpandi, J. Subramaniam, R. Rakesh, P. Amuthavalli, C. Panneerselvam, U. Suresh, C. Vadivalagan, M.S. Alsalhi, S. Devanesan, H. Wei, A. Higuchi, M. Nicoletti, A. Canale, G. Benelli, Mangrove helps: *Sonneratia alba*-synthesized silver nanoparticles magnify guppy fish predation against *Aedes aegypti* young instars and down-regulate the expression of envelope (E) gene in dengue virus (Serotype DEN-2). J. Clust. Sci. **28**, 437–461 (2017)

K. Murugan, C. Panneerselvam, J. Subramaniam, P. Madhiyazhagan, J.-S. Hwang, L. Wang, D. Dinesh, U. Suresh, M. Roni, A. Higuchi, M. Nicoletti, G. Benelli, Eco-friendly drugs from the marine environment: spongeweed-synthesized silver nanoparticles are highly effective on *Plasmodium falciparum* and its vector *Anopheles stephensi*, with little non-target effects on predatory copepods. Environ. Sci. Pollut. Res. **23**, 16671–16685 (2016b)

K. Murugan, M. Roni, C. Panneerselvam, A.T. Aziz, U. Suresh, R. Rajaganesh, R. Aruliah, J.A. Mahyoub, S. Trivedi, H. Rehman, H.A.N. Al-Aoh, S. Kumar, A. Higuchi, B. Vaseeharan, H.

Wei, S. Senthil-Nathan, A. Canale, G. Benelli, *Sargassum wightii*-synthesized ZnO nanoparticles reduce the fitness and reproduction of the malaria vector *Anopheles stephensi* and cotton bollworm *Helicoverpa armigera*. Physiol. Mol. Plant Pathol. **101**, 202–213 (2018)

K. Murugan, C.M. Samidoss, C. Panneerselvam, A. Higuchi, M. Roni, U. Suresh, B. Chandramohan, J. Subramaniam, P. Madhiyazhagan, D. Dinesh, R. Rajaganesh, A.A. Alarfaj, M. Nicoletti, S. Kumar, H. Wei, A. Canale, H. Mehlhorn, G. Benelli, Seaweed-synthesized silver nano- particles: an eco-friendly tool in the fight against *Plasmodium falciparum* and its vector *Anopheles stephensi*? Parasitol. Res. **114**, 4087–4097 (2015b)

S. Murugesan, S. Bhuvaneswari, N. Shanthi, P. Murugakoothan, V. Sivamurugan, Red alga *Hypnea musciformis* (Wulf) Lamour mediated environmentally benign synthesis and antifungal activity of gold nano particles. Int. J. NanoScience Nanotechnol. **6**, 71–83 (2015)

S. Murugesan, S. Bhuvaneswari, V. Sivamurugan, Green synthesis, characterization of silver nanoparticles of a marine red alga *Spyridia fusiformis* and their antibacterial activity. Int. J. Pharm. Pharm. Sci. **9**, 192–197 (2017)

S. Murugesan, M. Elumalai, R. Dhamotharan, Green synthesis of silver nano particles from marine alga *Gracilaria edulis*. Biosci. Biotech. Res. Comm. **4**, 105–110 (2011)

M.S. Muthu, P.P.J. John, C.U. Iniya, Green synthesis of silver nanoparticles using *Enteromorpha linza* (L.) J.AG. (green seaweed) from Hare island, Thoothukudi, Tamil Nadu, India. Eur. J. Mol. Biol. Biochem. **1**, 201–206 (2014)

A. Nabikhan, K. Kandasamy, A. Raj, N.M. Alikunhi, Synthesis of antimicrobial silver nanoparticles by callus and leaf extracts from saltmarsh plant, *Sesuvium portulacastrum* L. Colloids Surf. B Biointerfaces **79**, 488–493 (2010)

M. Nag, D. Lahiri, S. Joshi, R.R. Ray, Evaluation of algal active compounds as potent antibiofilm agent. J. Basic Microbiol. **62**, 1098–1109 (2022)

P. Nagababu, U.V. Rao, Cost-effective green synthesis and characterization of silver nanoparticles from *Avicennia alba* blume leaves and their antibacterial activity. Asian J. Pharm. Clin. Res. **9**, 301–303 (2016)

P. Nagababu, V.U. Rao, Pharmacological assessment, green synthesis and characterization of silver nanoparticles of *Sonneratia apetala* Buch.-Ham. Leaves. J. Appl. Pharm. Sci. **7**, 175–182 (2017)

S. Nagarajan, K.A. Kuppusamy, Extracellular synthesis of zinc oxide nanoparticle using seaweeds of gulf of Mannar, India. J. Nanobiotechnol. **11**, 39 (2013)

K.S.B. Naidu, N. Murugan, J.K. Adam, Sershen, Biogenic synthesis of silver nanoparticles from *Avicennia marina* seed extract and its antibacterial potential. Bionanoscience **9**, 266–273 (2019)

F. Namvar, S. Azizi, M.B. Ahmad, K. Shameli, R. Mohamad, M. Mahdavi, P.M. Tahir, Green synthesis and characterization of gold nanoparticles using the marine macroalgae *Sargassum muticum*. Res. Chem. Intermed. **41**, 5723–5730 (2015a)

F. Namvar, H.S. Rahman, R. Mohamad, A. Rasedee, S.K. Yeap, M.S. Chartrand, S. Azizi, P.M. Tahir, Apoptosis induction in human Leukemia cell lines by gold nanoparticles synthesized using the green biosynthetic approach. J. Nanomater. **2015**, 642621 (2015b)

B.E. Naveena, S. Prakash, Biological synthesis of gold nanoparticles using marine algae *Gracilaria corticata* and its application as a potent antimicrobial and antioxidant. Asian J. Pharm. Clin. Res. **6**, 179–182 (2013)

S. Naveenkumar, C. Kamaraj, C. Ragavendran, M. Vaithiyalingam, V. Sugumar, K. Marimuthu, Gracilaria corticata red seaweed mediate biosynthesis of silver nanoparticles: larvicidal, neurotoxicity, molecular docking analysis, and ecofriendly approach. Biomass Convers Biorefin (2023). https://doi.org/10.1007/s13399-023-04026-x

M.A. Negm, H.A.H. Ibrahim, N.A. Shaltout, H.A. Shawky, M.S. Abdel-mottaleb, S.K. Hamdona, Green synthesis of silver nanoparticles using marine algae extract and their antibacterial activity. Middle East J. Appl. Sci. **8**, 957–970 (2018)

C. Oliveira, N.M. Neves, R.L. Reis, A. Martins, T.H. Silva, Gemcitabine delivered by fucoidan/chitosan nanoparticles presents increased toxicity over human breast cancer cells. Nanomedicine **13**, 2037–2050 (2018)

H.H. Omar, F.S. Bahabri, A.M. El-Gendy, Biopotential application of synthesis nanoparticles as antimicrobial agents by using *Laurencia papillosa*. Int. J. Pharmacol. **13**, 303–312 (2017)

F.L. Oscar, S. Vismaya, M. Arunkumar, N. Thajuddin, D. Dhanasekaran, C. Nithya, Algal nanoparticles: synthesis and biotechnological potentials, in *Algae—Organisms for Imminent Biotechnology* ed. by N. Thajuddin, D. Dhanasekaran (IntechOpen, London, United Kingdom, 2016) pp. 157–182.

G. Oza, S. Pandey, R. Shah, M. Sharon, A mechanistic approach for biological fabrication of crystalline gold nanoparticles using Marine Algae, *Sargassum wightii*. Pelagia Res. Libr. **2**, 505–512 (2012)

B.Y. Öztürk, B.Y. Gürsu, İ Dağ, Antibiofilm and antimicrobial activities of green synthesized silver nanoparticles using marine red algae *Gelidium corneum*. Process Biochem. **89**, 208–219 (2020)

P. Palaniappan, G. Sathishkumar, R. Sankar, Fabrication of nano-silver particles using *Cymodocea serrulata* and its cytotoxicity effect against human lung cancer A549 cells line. Spectrochim. Acta A Mol. Biomol. Spectrosc. **138**, 885–890 (2015)

S. Palanisamy, P. Rajasekar, G. Vijayaprasath, G. Ravi, R. Manikandan, N.M. Prabhu, A green route to synthesis silver nanoparticles using *Sargassum polycystum* and its antioxidant and cytotoxic effects: an in vitro analysis. Mater. Lett. **189**, 196–200 (2017)

T. Palaniyandi, S. Viswanathan, P. Prabhakaran, G. Baskar, M.R.A. Wahab, A. Sivaji, M. Ravi, B.K. Rajendran, M. Moovendhan, H. Surendran, S. Kumarasamy, Green synthesis of gold nanoparticles using *Halymenia pseudofloresii* extracts and their antioxidant, antimicrobial, and anti-cancer activities. Biomass Conv. Bioref. (2023)

D. Parial, H.K. Patra, A.K.R. Dasgupta, R. Pal, Screening of different algae for green synthesis of gold nanoparticles. Eur. J. Phycol. **47**, 22–29 (2012)

S.Y. Park, Y.J. Kim, G. Park, H.-H. Kim, Neuroprotective effect of *Dictyopteris divaricata* extract-capped gold T nanoparticles against oxygen and glucose deprivation/reoxygenation. Colloids Surf. B Biointerfaces **179**, 421–428 (2019)

H.L. Parker, J.R. Dodson, V.L. Budarin, J.H. Clark, A.J. Hunt, Direct synthesis of Pd nanoparticles on alginic acid and seaweed supports. Green Chem. **17**, 2200 (2015)

R. Parthasarathy, S.P. Kumar, H.C.Y. Rao, J. Chelliah, Synthesis of β-glucan nanoparticles from red algae–derived β-glucan for potential biomedical applications. Appl. Biochem. Biotechnol. **193**, 3983–3995 (2021)

J.J.P. Paul, S.D.K.S. Devi, Biosynthesis and characterization of Silver nanoparticles using *Gracilaria dura* (AG.) J.AG. (Red Seaweed). Am. J. Pharmtech Res. **4**, 489–498 (2014)

V.K. Pawar, Y. Singh, K. Sharma, A. Shrivastav, A. Sharma, A. Singh, J.G. Meher, P. Singh, K. Raval, A. Kumar, H.K. Bora, D. Datta, J. Lal, M.K. Chourasia, Improved chemotherapy against breast cancer through immunotherapeutic activity of fucoidan decorated electrostatically assembled nanoparticles bearing doxorubicin. Int. J. Biol. Macromol. **122**, 1100–1114 (2019)

J.L. Pérez-Lloréns, A.T. Critchley, M.L. Cornish, O.G. Mouritsen, Saved by seaweeds (II): traditional knowledge, home remedies, medicine, surgery, and pharmacopoeia. J. Appl. Phycol. **35**, 2049–2068 (2023)

P. Pisitsak, K. Chamchoy, V. Chinprateep, W. Khobthong, P. Chitichotpanya, S. Ummartyotin, Synthesis of gold nanoparticles using tannin-rich extract and coating onto cotton textiles for catalytic degradation of congo red. J. Nanotechnol. **2021**, 6380283 (2021)

S. Poornima, K. Valivittan, Degradation of Malachite Green (Dye) by using photo-catalytic biogenic silver nanoparticles synthesized using Red Algae (*Gracilaria corticata*) aqueous extract. Int. J. Curr. Microbiol. App. Sci. **6**, 62–70 (2017)

D.K. Poudel, P. Niraula, H. Aryal, B. Budhathoki, S. Phuyal, R. Marahatha, K. Subedi, Plant-mediated green synthesis of AgNPs and their possible applications: a critical review. J. Nanotechnol. **2022**, 2779237 (2022)

B.S.N. Prasad, T.V.N. Padmesh, Seaweed (*Sargassum ilicifolium*) assisted green synthesis of palladium nanoparticles. Int. J. Sci. Eng. Res. **5**, 229–231 (2014)

T.N.V.K.V. Prasad, E.K. Elumalai, Marine algae mediated synthesis of silver nanopaticles using *Scaberia agardhii* Greville. J. Biol. Sci. **13**, 566–569 (2013)

T.N.V.K.V. Prasad, V.S.R. Kambala, R. Naidu, Phyconanotechnology: synthesis of silver nanoparticles using brown marine algae *Cystophora moniliformis* and their characterisation. J. Appl. Phycol. **25**, 177–182 (2013)

M. Premanathan, V.S.S. Benitha, K. Jeyasubramanian, K. Kathiresan, Rapid biosynthesis of antibacterial silver nanoparticles by *Rhizophora mucronata* Leaf. Adv. Sci. Eng. Med. **6**, 184–187 (2014)

K.F. Princy, A. Gopinath, Eco-friendly synthesis and characterization of silver nanoparticles using marine macroalga *Padina tetrastromatica*. Int. J. Sci. Res. **4**, 1050–1054 (2015)

R.I. Priyadharshini, G. Prasannaraj, N. Geetha, P. Venkatachalam, Microwave-mediated extracellular synthesis of metallic silver and zinc oxide nanoparticles using macro-algae (*Gracilaria edulis*) extracts and its anticancer activity against human PC3 cell lines. Appl. Biochem. Biotechnol. **174**, 2777–2790 (2014)

A. Pugazhendhi, D. Prabakar, J.M. Jacob, I. Karuppusamy, R.G. Saratale, Synthesis and characterization of silver nanoparticles using *Gelidium amansii* and its antimicrobial property against various pathogenic bacteri. Microb. Pathog. **114**, 41–45 (2018)

A. Pugazhendhi, R. Prabhu, K.M. Uruganantham, R. Shanmuganathan, S. Natarajan, Anticancer, antimicrobial and photocatalytic activities of green synthesized magnesium oxide nanoparticles (MgONPs) using aqueous extract of *Sargassum wightii*. J. Photochem. Photobiol. B **190**, 86–97 (2019)

M. Puskulluoglu, I. Michalak, An ocean of possibilities: a review of marine organisms as sources of nanoparticles for cancer care. Nanomedicine **17**, 1695–1719 (2022)

R.K. Raajshree, D. Brindha, In vivo anticancer activity of biosynthesized Zinc Oxide nanoparticle using *Turbinaria conoides* on a Dalton's lymphoma ascites mice model. J. Environ. Pathol. Toxicol. Oncol. **37**, 103–115 (2018)

R.S.R. Radhika, S. Gayathri, M. Gobalakrishnan, *Marine Biomolecule Mediated Synthesis of Selenium Nanoparticles and their Antimicrobial Efficiency Against Fish and Crustacean Pathogens* (2022). https://doi.org/10.21203/rs.3.rs-1594624/v1

Z. Rahimi, M. Yousefzadi, A. Noori, A. Akbarzadeh, Green synthesis of silver nanoparticles using *Ulva flexousa* from the Persian Gulf, Iran. J. Persian Gulf (Mar. Sci.) **5**, 9–16 (2014)

S. Rajaboopathi, S. Thambidurai, Green synthesis of seaweed surfactant based CdO-ZnO nanoparticles for better thermal and photocatalytic activit. Curr. Appl. Phys. **17**, 1622–1638 (2017)

P. Rajasulochana, R. Dhamotharan, P. Murugakoothan, S. Murugesan, P. Krishnamoorthy, Biosynthesis and characterization of gold nanoparticles using the alga *Kappaphycus alvarezii*. Int. J. Nanosci. **9**, 511–516 (2010)

P. Rajasulochana, P. Krishnamoorthy, R. Dhamotharan, Potential application of *Kappaphycus alvarezii* in agricultural and pharmaceutical industry. J. Chem. Pharm. Res. **4**, 33–37 (2012)

F.A.A. Rajathi, C. Parthiban, V.G. Kumar, P. Anantharaman, Biosynthesis of antibacterial gold nanoparticles using brown alga, *Stoechospermum marginatum* (kutzing). Spectrochim. Acta A Mol. Biomol. Spectrosc. **99**, 166–173 (2012)

S. Rajesh, D.P. Raja, J.M. Rathi, K. Sahayaraj, Biosynthesis of silver nanoparticles using *Ulva fasciata* (Delile) ethyl acetate extract and its activity against *Xanthomonas campestris* pv. *malvacearum*. Jbiopest **5**, 119–128 (2012)

S. Rajeshkumar, Synthesis of zinc oxide nanoparticles using algal formulation (*Padina tetrastromatica* and *Turbinaria conoides*) and their antibacterial activity against fish pathogens. Res. J. Biotechnol. **13**, 15–19 (2018)

S. Rajeshkumar, Phytochemical constituents of fucoidan (*Padina tetrastromatica*) and its assisted AgNPs for enhanced antibacterial activity. IET Nanobiotechnol. **11**, 292–299 (2017)

S. Rajeshkumar, C. Kannan, G. Annadurai, Synthesis and characterization of antimicrobial silver nanoparticles using marine brown seaweed *Padina tetrastromatica*. Drug Invention Today **4**, 511–513 (2012a)

S. Rajeshkumar, C. Kannan, G. Annadurai, Green synthesis of silver nanoparticles using marine brown algae *Turbinaria conoides* and its antibacterial activity. Int. J. Pharm. Bio. Sci. **3**, 502–510 (2012b)

S. Rajeshkumar, S.V. Kumar, C. Malarkodi, M. Vanaja, K. Paulkumar, G. Annadurai, Optimized synthesis of gold nanoparticles using green chemical process and its invitro anticancer activity against HepG2 and A549 cell Lines4. Mech. Mater. Sci. Eng. **9**, 1–7 (2017a)

S. Rajeshkumar, C. Malarkodi, G. Gnanajobitha, K. Paulkumar, M. Vanaja, C. Kannan, G. Annadurai, Seaweed-mediated synthesis of gold nanoparticles using *Turbinaria conoides* and its characterization. J. Nanostructure Chem. **3**, 44 (2013a)

S. Rajeshkumar, C. Malarkodi, V.S. Kumar, Synthesis and characterization of silver nanoparticles from marine brown seaweed and its antifungal efficiency against clinical fungal pathogens. Asian J. Pharm. Clin. Res. **10**, 190–193 (2017b)

S. Rajeshkumar, C. Malarkodi, K. Paulkumar, M. Vanaja, G. Gnanajobitha, G. Annadurai, Algae mediated green fabrication of silver nanoparticles and examination of its antifungal activity against clinical pathogens. Int. J. Metals **2014**, 692643 (2014)

S. Rajeshkumar, C. Malarkodi, M. Vanaja, G. Gnanajobitha, K. Paulkumar, G. Annadura, C. Kannan, Antibacterial activity of algae mediated synthesis of gold nanoparticles from *Turbinaria conoides*. Der Pharma Chemica **5**, 224–229 (2013b)

S. Rajeshkumar, N.T. Nandhini, K. Manjunath, P. Sivaperumal, G.K. Prasad, S.S. Alotaibi, S.M. Roopan, Environment friendly synthesis copper oxide nanoparticles and its antioxidant, antibacterial activities using Seaweed (*Sargassum longifolium*) extract. J. Mol. Struct. **1242**, 130724 (2021)

B. Raju, A. Muniyasamy, S.G. Prakash, A.S. Sundararaj, U. Kesavachandran, Phycosynthesis of nanostructured Silver using *Enteromorpha intestinalis* and evaluation of its inhibitory effect on human bacterial and fungal pathogens. J. Clust. Sci. **28**, 1739–1748 (2017)

V.P. Ram, J. Yasur, P. Abishad, V. Unni, D.P. Gourkhede, M.A.D. Nishanth, P. Niveditha, J. Vergis, S.V.S. Malik, K. Byrappa, N.V. Kurkure, D.B. Rawool, S.B. Barbuddhe, Antimicrobial efficacy of green synthesized nanosilver-conjugated cinnamaldehyde against multi-drug-resistant enteroaggregative *Escherichia coli* in *Galleria mellonella*. Pharmaceutics **14**, 1924 (2022)

M. Ramakrishna, D.R. Babu, R.M. Gengan, S. Chandra, G.N. Rao, Green synthesis of gold nanoparticles using marine algae and evaluation of their catalytic activity. J. Nanostruct. Chem. **6**, 1–13 (2016)

C.M. Ramakritinan, E. Kaarunya, S. Shankar, A.K. Kumaraguru, Antibacterial effects of Ag, Au and bimetallic (Ag-Au) nanoparticles synthesized from red algae. Solid State Phenom. **201**, 211–230 (2013)

N. Ramalingam, C. Rose, C. Krishnan, S. Sankar, Green synthesis of silver nanoparticles using red marine algae and evaluation of its antibacterial activity. J. Pharm. Sci. Res. **10**, 2435–2438 (2018)

R. Ramamoorthy, S. Vanitha, P. Krishnadev, Green synthesis of silver nanoparticles using red seaweed *Portieria hornemannii* (Lyngbye) P.C. silva and its antifungal activity against silkworm (*Bombyx mori* L.) Muscardine pathogens. J. Pharmacogn. Phytochem. **8**, 3394–3398 (2019)

S.V.P. Ramaswamy, S. Narendhran, R. Sivaraj, Potentiating effect of ecofriendly synthesis of copper oxide nanoparticles using brown alga: antimicrobial and anticancer activities. Bull. Mater. Sci. **39**, 361–364 (2016)

C.H. Ramesh, S. Koushik, T. Shunmugaraj, M.V.R. Murthy, Occurrence of green tides on Palk Bay and Gulf of Mannar regions, Southeast coast of Tamil Nadu, India. J. Mar. Biol. Assoc. India (2020)

V.S. Ramkumar, A. Pugazhendhi, K. Gopalakrishnan, P. Sivagurunathan, G.D. Saratale, T.N.B. Dung, E. Kannapiran, Biofabrication and characterization of silver nanoparticles using aqueous extract of seaweed *Enteromorpha compressa* and its biomedical properties. Biotechnol. Rep. **14**, 1–7 (2017a)

V.S. Ramkumar, A. Pugazhendhi, S. Prakash, N.K. Ahila, G. Vinoj, S. Selvam, G. Kumar, E. Kannapiran, R.B. Rajendran, Synthesis of platinum nanoparticles using seaweed *Padina gymnospora* and their catalytic activity as PVP/PtNPs nanocomposite towards biological applications. Biomed. Pharmacother. **92**, 479–490 (2017b)

L. Ramteke, B.L. Jadhav, P. Gawali, Biogenic copper nanoparticles from the aqueous stem extract of *Ceriops tagal*. World J. Pharm. Res. **7**, 933–947 (2018)

P. RathnaKumari, P. Kolanchinathan, D. Siva, B. Abirami, V. Masilamani, G. John, S. Achiraman, A. Balasundaram, Antibacterial efficacy of seagrass *Cymodocea serrulata*-engineered silver nanoparticles against prawn pathogen Vibrio parahaemolyticus and its combative effect on the marine shrimp Penaeus monodon. Aquaculture **493**, 158–164 (2018)

A. Ravichandran, P. Subramanian, V. Manoharan, T. Muthu, R. Periyannan, M. Thangapandi, K. Ponnuchamy, B. Pandi, P.N. Marimuthu, Phyto-mediated synthesis of silver nanoparticles using fucoidan isolated from *Spatoglossum asperum* and assessment of antibacterial activities. J. Photochem. Photobiol. B **185**, 117–125 (2018)

R.R. Remya, S.R.R. Rajasree, L. Aranganathan, T.Y. Suman, S. Gayathri, Enhanced cytotoxic activity of AgNPs on retinoblastoma Y79 cell lines synthesised using marine seaweed *Turbinaria ornata*. IET Nanobiotechnol. **11**, 18–23 (2017)

A. Rojas-Pérez, L. Adorno, M. Cordero, A. Ruiz, Z. Mercado-Díaz, A. Rodríguez, L. Betancourt, C. Vélez, I. Feliciano, C. Cabrera, L.M. Díaz-Vázquez, Biosynthesis of gold nanoparticles using *Osmundaria obtusiloba* extract and their potential use in optical sensing applications. Austin J. Biosensors Bioelectronics **1**, 1014 (2015)

M. Roni, K. Murugan, C. Panneerselvam, J. Subramaniam, M. Nicoletti, P. Madhiyazhagan, D. Dinesh, U. Suresh, H.F. Khater, H. Wei, A. Canale, A.A. Alarfaj, M.A. Munusamy, A. Higuchi, G. Benelli, Characterization and biotoxicity of *Hypnea musciformis* synthesized silver nanoparticles as potential eco-friendly control tool against *Aedes aegypti* and *Plutella xylostella*. Ecotoxicol. Environ. Saf. **121**, 31–38 (2015)

T.A. Roseline, M. Murugan, M.P. Sudhakar, K. Arunkumar, Nanopesticidal potential of silver nanocomposites synthesized from the aqueous extracts of red seaweeds. Environ. Technol. Innov. **13**, 82–93 (2019)

S. Roy, A review: green synthesis of nanoparticles from seaweeds and its some applications. Austin J. Nanomed. Nanotechnol. **7**, 1054 (2019)

S. Roy, P. Anantharaman, Biosynthesis of silver nanoparticles by *Chlorodesmis hildebrandtii* A. Gepp & E. Gepp including its agricultural and biomedical implications. Nanomed. Nanotechnol. Open Access **3**, 000144 (2018a)

S. Roy, P. Anantharaman, Agricultural and biomedical application of silver nanoparticles synthesized by *Halimeda gracilis* Harvey ex J. Agardh. Int. J. Environ. Agric. Biotechnol. **3**, 1739–1747 (2018b)

S. Roy, P. Anantharaman, Biosynthesis of silver nanoparticles by *Sargassum ilicifolium* (Turner) C. Agardh with their antimicrobial activity and potential for seed germination. J. Appl. Phys. Nanotechnol. **1**, 002 (2018)

S. Roy, P. Anantharaman, Biosynthesis of silver nanoparticle by *Amphiroa anceps* (Lamarck) decaisne and its biomedical and ecological implications. J. Nanomed. Nanotechnol. **9**, 1000492 (2018d)

S. Roy, P. Anantharaman, Biosynthesis of silver nanoparticles by *Chaetomorpha antennina* (Bory de Saint-Vincent) Kutzing with its antibacterial activity and ecological implication. J. Nanomed. Nanotechnol. **8**, 5 (2017)

P. Sabatini, C. Anchana Devi, Green synthesis of silver nanoparticles from red seaweed (*Portieria hornemannii*) and its applications. Eur. J. Biomed. Pharmaceutical Sci. **4**, 949–956 (2017)

H. Saber, E.A. Alwaleed, K.A. Ebnalwaled, A. Sayed, W. Salem, Efficacy of silver nanoparticles mediated by *Jania rubens* and *Sargassum dentifolium* macroalgae; Characterization and biomedical applications. Egyptian J. Basic Appl. Sci. **4**, 249–255 (2017)

M. Safaat, S. Tursiloadi, B. Perisha, F. Zulpikar, Nanoparticles green synthesis macroalgae-based and its application and distribution in Indonesia—an overview. Int. Symp. Aquatic Sci. Resources Manage. **744**, 012067 (2021)

K. Sahayaraj, S. Rajesh, J.A.M. Rathi, V. Kumar, Green preparation of seaweed-based silver nano-liquid for cotton pathogenic fungi management. IET Nanobiotechnol. **13**, 219–225 (2018)

K. Sahayaraj, S. Rajesh, J.M. Rathi, Silver nanoparticles biosynthesis using marine alga *Padina pavonica* (Linn.) and its microbicidal activity. Dig. J. Nanomaterials Biostructures **7**, 1557–1567 (2012)

D.M.S.A. Salem, M.M. Ismail, M.A. Aly-Eldeen, Biogenic synthesis and antimicrobial potency of iron oxide (Fe_3O_4) nanoparticles using algae harvested from the Mediterranean Sea, Egypt. Egypt J. Aquat. Res. **45**, 197–204 (2019)

D.M.S.A. Salem, M.M. Ismail, H.R.Z. Tadros, Evaluation of the antibiofilm activity of three seaweed species and their biosynthesized iron oxide nanoparticles (Fe_3O_4-NPs). Egypt. J. Aquat. Res. **46**, 333–339 (2020)

S.S. Salem, A. Fouda, Green synthesis of metallic nanoparticles and their prospective biotechnological applications: an overview. Biol. Trace Elem. Res. **199**, 344–370 (2021)

Z. Sanaeimehr, I. Javadi, F. Namvar, Antiangiogenic and antiapoptotic effects of green-synthesized zinc oxide nanoparticles using *Sargassum muticum* algae extraction. Cancer Nanotechnol. **9**, 3 (2018)

N. Sangeetha, S. Manikandan, M. Singh, A.K. Kumaraguru, Biosynthesis and characterization of silver nanoparticles using freshly extracted sodium alginate from the seaweed *Padina tetrastromatica* of Gulf of Mannar, India. Curr. Nanosci. **8**, 1–6 (2012)

M.V. Sankar, S. Abideen, Biosynthesis of silver nanoparticles using *Rhizophora mucronata* and *Ceriops decandra* and their antagonistic activity on gut cellulolytic bacteria. Int. J. Pharmaceutical Biol. Arch. **10**, 138–145 (2019)

R.N. Sari, Nurhasni, M.A. Yaqin, Sintesis nanopartikel zno ekstrak *Sargassum* sp. Dan karakteristik produknya. J. Pengolah Has Perikan Indones **20**, 238–254 (2017)

T. Sathyaseelan, M. Subbiah, V. Sivamugugan, Green synthesis of silver nano particles using marine brown alga *Lobophora variegata* and its efficacy in antifungal activity. World J. Pharm. Res. **4**, 2137–2145 (2015)

L. Satish, S. Santhakumari, S. Gowrishankar, S.K. Pandian, A.V. Ravi, M. Ramesh, Rapid biosynthesized AgNPs from *Gelidiella acerosa* aqueous extract mitigates quorum sensing mediated biofilm formation of *Vibrio* species—an in vitro and in vivo approach. Environ. Sci. Pollut. Res. **24**, 27254–27268 (2017)

K. Satyavani, S. Gurudeeban, T. Ramanathan, T. Balasubramanian, *Ipomoea pes-caprae* Mediated silver nanoparticles and their antibacterial effect. Sci. Int. **1**, 155–159 (2013)

K. Satyavani, S. Gurudeeban, T. Ramanathan, T. Balasubramanian, Toxicity study of silver nanoparticles synthesized from *Suaeda monoica* on Hep-2 cell line. Avicenna J. Med. Biotechnol. **4**, 35–39 (2012)

K. Satyavani, S. Gurudeeban, T. Ramanathan, T. Balasubramanian, Biomedical potential of silver nanoparticles synthesized from calli cells of *Citrullus colocynthis* (L.) Schrad. J. Nanobiotechnol. **9**, 43 (2011)

G.G. Selvam, K. Sivakumar, Phycosynthesis of silver nanoparticles and photocatalytic degradation of methyl orange dye using silver (Ag) nanoparticles synthesized from *Hypnea musciformis* (Wulfen) J.V. Lamouroux. Appl. Nanosci. **5**, 617–622 (2015)

G.G. Selvam, K. Sivakumar, Phycosynthesis and photocatalytic degradation of methyl orange using silver nanoparticles synthesized by the *Sargassum wightii*. World J. Pharm. Sci. **2**, 1022–1028 (2014)

P. Selvaraj, E. Neethu, P. Rathika, J.P.R. Jayaseeli, B.R. Jermy, S. AbdulAzeez, J.F. Borgio, T.S. Dhas, Antibacterial potentials of methanolic extract and silver nanoparticles from marine algae. Biocatal. Agric. Biotechnol. **28**, 101719 (2020)

B.C.G. Selvi, J. Madhavan, A. Santhanam, Cytotoxic effect of silver nanoparticles synthesized from *Padina tetrastromatica* on breast cancer cell line. Adv. Natural Sci. Nanosci. Nanotechnol. **7**, 035015 (2016)

P. Senthilkumar, D.S.R.S. Kumar, B. Sudhagar, M. Vanthana, M.H. Parveen, S. Sarathkumar, J.C. Thomas, A.S. Mary, C. Kannan, Seagrass-mediated silver nanoparticles synthesis by *Enhalus acoroides* and its α-glucosidase inhibitory activity from the Gulf of Mannar. J. Nanostruct. Chem. **6**, 275–280 (2016)

P. Senthilkumar, L. Surendran, B. Sudhagar, D.S.R.S. Kumar, Facile green synthesis of gold nanoparticles from marine algae *Gelidiella acerosa* and evaluation of its biological potential. SN Appl. Sci. **1**, 284 (2019)

A.M.E. Shafey, Green synthesis of metal and metal oxide nanoparticles from plant leaf extracts and their applications: a review. Green Process. Synthesis **9**, 304–339 (2020)

K. Shahzamani, H.E. Lashgarian, M. Karkhane, A. Ghaffarizadeh, S. Ghotekar, A. Marzban, Bioactivity assessments of phyco-assisted synthesized selenium nanoparticles by aqueous extract of green seaweed, *Ulva fasciata*. Emergent Mater. **5**, 1689–1698 (2022)

N. Shanmugam, P. Rajkamal, S. Cholan, N. Kannadasan, K. Sathishkumar, G. Viruthagiri, A. Sundaramanickam, Biosynthesis of silver nanoparticles from the marine seaweed *Sargassum wightii* and their antibacterial activity against some human pathogens. Appl. Nanosci. **4**, 881–888 (2014)

B. Sharma, D.D. Purkayastha, S. Hazra, M. Thajamanbi, C.R. Bhattacharjee, N.N. Ghosh, J. Rout, Biosynthesis of fluorescent gold nanoparticles using an edible freshwater red alga, *Lemanea fluviatilis* (L.) C.Ag. and antioxidant activity of biomatrix loaded nanoparticles. Bioprocess Biosyst. Eng. **37**, 2559–2565 (2014)

P.J. Shiny, S.P. Dhas, A. Mukherjee, N. Chandrasekaran, *Padina tetrastomatica* : a potential source for the synthesis of silver nanoparticles and its antibacterial efficiency. Adv. Sci. Eng. Med. **5**, 926–931 (2013a)

P.J. Shiny, A. Mukherjee, N. Chandrasekaran, Haemocompatibility assessment of synthesised platinum nanoparticles and its implication in biology. Bioprocess Biosyst. Eng. **37**, 991–997 (2014)

P.J. Shiny, A. Mukherjee, N. Chandrasekaran, Marine algae mediated synthesis of the silver nanoparticles and its antibacterial efficiency. Int. J. Pharm. Pharm. Sci. **5**, 239–241 (2013b)

M.K. Shukla, R.P. Singh, C.R.K. Reddy, B. Jha, Synthesis and characterization of agar-based silver nanoparticles and nanocomposite film with antibacterial applications. Bioresour. Technol. **107**, 295–300 (2012)

G. Singaravelu, J.S. Arockiamary, V.G. Kumar, K. Govindaraju, A novel extracellular synthesis of monodisperse gold nanoparticles using marine alga, *Sargassum wightii* Greville. Colloids Surf. B Biointerfaces **57**, 97–101 (2007)

J. Singh, T. Dutta, K. Kim, M. Rawat, P. Samddar, P. Kumar, 'Green' synthesis of metals and their oxide nanoparticles: applications for environmental remediation. J. Nanobiotechnol. **16**, 84 (2018)

M. Singh, R. Kalaivani, S. Manikandan, N. Sangeetha, A.K. Kumaraguru, Facile green synthesis of variable metallic gold nanoparticle using *Padina gymnospora*, a brown marine macroalga. Appl. Nanosci. **3**, 145–151 (2013a)

M. Singh, M. Kumar, R. Kalaivani, S. Manikandan, A.K. Kumaraguru, Metallic silver nanoparticle: a therapeutic agent in combination with antifungal drug against human fungal pathogen. Bioprocess Biosyst. Eng. **36**, 407–415 (2013b)

M. Singh, M. Kumar, S. Manikandan, N. Chandrasekaran, A. Mukherjee, A. Kumaraguru, Drug delivery system for controlled cancer therapy using physico-chemically stabilized bioconjugated gold nanoparticles synthesized from Marine Macroalgae, *Padina gymnospora*. Nanomed. Nanotechnol. **S:5**, 1–7 (2014)

M. Singh, K. Saurav, M. Kumar, M. Kumari, S. Manikandan, A.K. Kumaraguru, The cytotoxicity and cellular stress by temperature fabricated polyshaped gold nanoparticles, using marine macroalgae, *Padina gymnospora*. Biotechnol. Appl. Biochem. **62**, 424–432 (2015)

S.P. Sivagnanam, A.T. Getachew, J.H. Choi, Y.B. Park, H.C. Woo, B.S. Chun, Green synthesis of silver nanoparticles from deoiled brown algal extract via Box-Behnken based design and their antimicrobial and sensing properties. Green Process Synth. **6**, 147–160 (2017)

R. Sivaraj, S.V.R. Priya, P. Rajiv, V. Rajendran, *Sargassum polycystum* C. Agardh mediated synthesis of gold nanoparticles assessing its characteristics and its activity against water borne pathogens. J. Nanomed. Nanotechnol. **6**, 1000280 (2015)

S. Soisuwan, W. Warisnoicharoen, K. Lirdprapamongkol, J. Svasti, Eco-friendly synthesis of fucoidan-stabilized gold nanoparticles. Am. J. Appl. Sci. **7**, 1038–1042 (2010)

H. Sonbol, F. Ameen, S. AlYahya, A. Almansob, S. Alwakeel, *Padina boryana* mediated green synthesis of crystalline palladium nanoparticles as potential nanodrug against multidrug resistant bacteria and cancer cells. Sci. Rep. **11**, 5444 (2021)

V.S.P.S. Sri, Y.A.S. Kumar, M. Savurirajan, D.K. Jha, N.V. Vinithkumar, G. Dharani, Anticancer efficacy of magnetite nanoparticles synthesized using aqueous extract of brown seaweed Rosenvingea intricata, South Andaman, India. Sci. Rep. **14**, 20255 (2024)

M. Subbiah, M. Pandithurai, S. Vajiravelu, *Spatoglossum asperum* J. Agardh mediated synthesis of silver nanoparticles, characterization and evaluation antifungal activities. J Pharmacogn. Phytochem. **8**, 1991–1995 (2019)

A. Subbulakshmi, S. Durgadevi, S. Anitha, M. Govarthanan, M. Biruntha, P. Rameshthangam, P. Kumar, Biogenic gold nanoparticles from *Gelidiella acerosa*: bactericidal and photocatalytic degradation of two commercial dyes. Appl. Nanosci. **13**, 4033–4042 (2022)

V. Subha, R.R.S. Ernest, P. Sruthi, S. Renganathan, An eco-friendly approach for synthesis of silver nanoparticles using *Ipomoea pes-caprae* root extract and their antimicrobial properties. Asian J. Pharm. Clin. Res. **8**, 103–106 (2015)

G. Sudha, A. Balasundaram, Synthesis and characterization of silver nanoparticles using *Padina pavonica* extract and evaluation of their antibacterial activity. J. Nanosci. Technol. **4**, 424–426 (2018)

M.P. Sudhakar, S. Venkatnarayanan, G. Dharani, Fabrication and characterization of bio-nanocomposite films using κ-Carrageenan and Kappaphycus alvarezii seaweed for multiple industrial applications. Int. J. Biol. Macromol. **219**, 138–149 (2022)

S. Sugandhi, G. Rani, Biosynthesis and characterization of gold nanoparticles from *Gracilaria corticata*. Nano Sci. Nano Technol. **8**, 475–481 (2014)

S. Suganya, B. Dhanalakshmi, S.D. Kumar, Biosynthesis and characterization of silver nanoparticles from *Sargassum wightii* and its antibacterial activity against multi-resistant human pathogens. Indian J. Geomarine Sci. **49**, 839–844 (2020)

V. Sujitha, K. Murugan, D. Dinesh, A. Pandiyan, R. Aruliah, J.-S. Hwang, K. Kalimuthu, C. Panneerselvam, A. Higuchi, A.T. Aziz, S. Kumar, A.A. Alarfaj, B. Vaseeharan, A. Canale, G. Benelli, Green-synthesized CdS nano-pesticides: toxicity on young instars of malaria vectors and impact on enzymatic activities of the non-target mud crab *Scylla serrata*. Aquat. Toxicol. **188**, 100–108 (2017)

S. Sunitha, A.N. Rao, L.S. Abraham, E. Dhayalan, R. Thirugnanasambandam, V.G. Kumar, Enhanced bactericidal effect of silver nanoparticles synthesized using marine brown macro algae. J. Chem. Pharm. Res. **7**, 191–195 (2015)

J. Suriya, R.S. Bharathi, V. Sekar, R. Rajasekaran, Biosynthesis of silver nanoparticles and its antibacterial activity using seaweed *Urospora* sp. Afr. J. Biotechnol. **11**, 12192–12198 (2012)

N. Thangaraju, R.P. Venkatalakshmi, A. Chinnasamy, P. Kannaiyan, Synthesis of silver nanoparticles and the antibacterial and anticancer activities of the crude extract of *Sargassum polycystum* C. Agardh. Nano Biomed. Eng. **4**, 89–94 (2012)

S. Thanigaivel, J. Thomas, A.S. Vickram, K. Anbarasu, R. Karunakaran, J. Palanivelu, P.S. Srikumar, Efficacy of encapsulated biogenic silver nanoparticles and its disease resistance against *Vibrio harveyi* through oral administration in *Macrobrachium rosenbergii*. Saudi J. Biol. Sci. **28**, 7281–7289 (2021)

P. Thatoi, R.G. Kerry, S. Gouda, G. Das, K. Pramanik, H. Thatoi, J.K. Patra, Photo-mediated green synthesis of silver and zinc oxide nanoparticles using aqueous extracts of two mangrove plant species, *Heritiera fomes* and *Sonneratia apetala* and investigation of their biomedical applications. J. Photochem. Photobiol. B **163**, 311–318 (2016)

R. Thiruchelvi, P. Jayashree, K. Mirunaalini, Synthesis of silver nanoparticle using marine red seaweed *Gelidiella acerosa*—a complete study on its biological activity and its characterisation. Mater. Today Proc. **37**, 1693–1698 (2021)

V.K. Thirumalairaj, M.P. Vijayan, G. Durairaj, L. Shanmugaasokan, R. Yesudas, S. Gunasekaran, Potential antibacterial activity of crude extracts and silver nanoparticles synthesized from *Sargassum wightii*. Int. Curr. Pharmaceutical J. **3**, 322–325 (2014)

R. Thiurunavukkarau, S. Shanmugam, K. Subramanian, P. Pandi, G. Muralitharan, M. Arokiarajan, K. Kasinathan, A. Sivaraj, R. Kalyanasundaram, S.Y. AlOmar, V. Shanmugam, Silver nanoparticles synthesized from the seaweed *Sargassum polycystum* and screening for their biological potential. Sci. Rep. **12**, 14757 (2022)

S. Tian, K. Saravanan, R.A. Mothana, G. Ramachandran, G. Rajivgandhi, N. Manoharan, Anticancer activity of biosynthesized silver nanoparticles using *Avicennia marina* against A549 lung cancer cells through ROS/mitochondrial damages. Saudi J. Biol. Sci. **27**, 3018–3024 (2020)

H.E. Touliabah, M.M. El-Sheekh, M.E.M. Makhlof, Evaluation of *Polycladia myrica* mediated selenium nanoparticles (PoSeNPS) cytotoxicity against PC-3 cells and antiviral activity against HAV HM175 (Hepatitis A), HSV-2 (Herpes simplex II), and Adenovirus strain 2. Front. Mar. Sci. **9**, 1092343 (2022)

S. Trivedi, M.A. Alshehri, A.T. Aziz, C. Panneerselvam, H.A. Al-Aoh, F. Maggi, S. Sut, S. Dall'Acqua, Insecticidal, antibacterial and dye adsorbent properties of *Sargassum muticum* decorated nano-silver particles. South Afr. J. Botany **139**, 432–441 (2021)

S. Ulagesan, T. Nam, Y. Choi, Biogenic preparation and characterization of *Pyropia yezoensis* silver nanoparticles (P.y AgNPs) and their antibacterial activity against *Pseudomonas aeruginosa*. Bioprocess Biosyst Eng **44**, 443–452 (2021)

J. Umashankari, D. Inbakandan, T.T. Ajithkumar, T. Balasubramanian, Mangrove plant, *Rhizophora mucronata* (Lamk, 1804) mediated one pot green synthesis of silver nanoparticles and its antibacterial activity against aquatic pathogens. Saline Syst. **8**, 11 (2012)

P. Usha Rani, K. Prasanna Laxmi, V. Vadlapudi, B. Sreedhar, Phytofabrication of silver nanoparticles using the mangrove associate, *Hibiscus tiliaceus* plant and its biological activity against certain insect and microbial pests. Jbiopest **9**, 167–179 (2016)

V. Vadlapudi, R. Amanchy, Synthesis, characterization and antibacterial activity of silver nanoparticles from red algae, *Hypnea musciformis*. Adv. Biol. Res. (Rennes) **11**, 242–249 (2017)

N. Valarmathi, F. Ameen, A. Almansob, P. Kumar, S. Arunprakash, M. Govarthanan, Utilization of marine seaweed *Spyridia filamentosa* for silver nanoparticles synthesis and its clinical applications. Mater. Lett. **263**, 127244 (2020)

S. Varun, S. Sudha, P.S. Kumar, Biosynthesis of gold nanoparticles from aqueous extract of *Dictyota bartayresiana* and their antifungal activity. Indian J. Adv. Chem. Sci. **2**, 190–193 (2014)

R.D. Vasquez, J.G. Apostol, J.D. de Leon, J.D. Mariano, C.M.C. Mirhan, S.S. Pangan, A.G.M. Reyes, E.T. Zamora, Polysaccharide-mediated green synthesis of silver nanoparticles from *Sargassum siliquosum* J.G. Agardh: Assessment of toxicity and hepatoprotective activity. OpenNano **1**, 16–24 (2016)

S. Veeramani, R R S. Ernest, R. Preethi, C. Joseph, K. Shanmugam, S. Renganathan, Silver nanoparticles—green synthesis with Aq. extract of stems *Ipomoea pes-Caprae*, characterization, antimicrobial and anti-cancer potential. Int. J. Med. Nano Res. **5**, 024 (2018)

C.K. Venil, C. Ramesh, P.R. Devi, L. Dufossé, Marine algal colorants for the food industry, in *Sustainable Global Resources of Seaweeds* ed. by A.R. Rao, G.A. Ravishankar, vol. 2 (Springer Nature, Switzerland AG, 2022), pp. 163–179

J. Venkatesan, S.-K. Kim, M.S. Shim, Antimicrobial, antioxidant, and anticancer activities of biosynthesized silver nanoparticles using Marine Algae *Ecklonia cava*. Nanomaterials **6**, 235 (2016)

J. Venkatesan, P. Manivasagan, S.-K. Kim, A.V. Kirthi, S. Marimuthu, A.A. Rahuman, Marine algae-mediated synthesis of gold nanoparticles using a novel *Ecklonia cava*. Bioprocess Biosyst. Eng. **37**, 1591–1597 (2014)

V. Venkatpurwar, V. Mali, S. Bodhankar, V. Pokharkar, In vitro cytotoxicity and in vivo sub-acute oral toxicity assessment of porphyran reduced gold nanoparticles. Toxicol. Environ. Chem. **94**, 1357–1367 (2012)

V. Venkatpurwar, V. Pokharkar, Green synthesis of silver nanoparticles using marine polysaccharide: study of in-vitro antibacterial activity. Mater. Lett. **65**, 999–1002 (2011)

V. Venkatpurwar, A. Shiras, V. Pokharkar, Porphyran capped gold nanoparticles as a novel carrier for delivery of anticancer drug: in vitro cytotoxicity study. Int. J. Pharm. **409**, 314–320 (2011)

A. Venkatraman, S.A.M. Yahoob, Y. Nagarajan, S. Harikrishnan, S. Vasudevan, T. Murugasamy, Pharmacological activity of biosynthesized gold nano-particles from brown algae-seaweed *Turbinaria conoides*. NanoWorld J. **4**, 17–22 (2018)

N. Vidyasagar, R.R. Patel, S.K. Singh, M. Singh, Green synthesis of silver nanoparticles: methods, biological applications, delivery and toxicity. Mater. Adv. **4**, 1831 (2023)

A.P. Vieira, E.M. Stein, D.X. Andreguetti, P. Colepicolo, A.M.C. Ferreira, Preparation of silver nanoparticles using aqueous extracts of the red algae *Laurencia aldingensis* and *Laurenciella* sp. and their cytotoxic activities. J. Appl. Phycol. **28**, 2615–2622 (2016)

S. Vijayakumar, J. Chen, V. Kalaiselvi, K. Tungare, M. Bhoric, Z.I. Gonzalez-Sanchez, E.F. Duran-Lara, Marine polysaccharide laminarin embedded ZnO nanoparticles and their based chitosan capped ZnO nanocomposites: synthesis, characterization and in vitro and in vivo toxicity assessment. Environ. Res. **213**, 113655 (2022)

S. Vijayakumar, K. Saravanakumar, B. Malaikozhundan, M. Divya, B. Vaseeharan, E.F. Durán-Lara, M.-H. Wang, Biopolymer K-carrageenan wrapped ZnO nanoparticles as drug delivery vehicles for anti MRSA therapy. Int. J. Biol. Macromol. **144**, 9–18 (2020)

S.R. Vijayan, P. Santhiyagu, M. Singamuthu, N.K. Ahila, R. Jayaraman, K. Ethiraj, Synthesis and characterization of silver and gold nanoparticles using aqueous extract of seaweed, *Turbinaria conoides*, and their antimicrofouling activity. Sci. World J. **2014**, 938272 (2014)

K. Vijayaraghavan, A. Mahadevan, M. Sathishkumar, S. Pavagadhi, R. Balasubramanian, Biosynthesis of Au(0) from Au(III) via biosorption and bioreduction using brown marine alga *Turbinaria conoides*. Chem. Eng. J. **167**, 223–227 (2011)

M. Vikneshan, R. Saravanakumar, R. Mangaiyarkarasi, S. Rajeshkumar, S.R. Samuel, M. Suganya, G. Baskar, Algal biomass as a source for novel oral nano-antimicrobial agent. Saudi J. Biol. Sci. **27**, 3753–3758 (2020)

S. Vinoth, S.G. Shankar, P. Gurusaravanan, B. Janani, J.K. Devi, Anti-larvicidal activity of silver nanoparticles synthesized from *Sargassum polycystum* against mosquito vectors. J. Clust. Sci. **30**, 171–180 (2019)

D. Vinu, K. Govindaraju, R. Vasantharaja, S.A. Nisa, M. Kannan, K.V. Anand, Biogenic zinc oxide, copper oxide and selenium nanoparticles: preparation, characterization and their anti-bacterial activity against *Vibrio parahaemolyticus*. J. Nanostructure Chem. **11**, 271–286 (2021)

K.M. Vishnu, S. Murugesan, In vitro cytotoxic activity of silver nano particle biosynthesized from *Colpomenia sinuosa* and *Halymenia poryphyroides* using DLA and EAC cell lines. World J. Pharm. Sci. **2**, 926–930 (2014)

S. Viswanathan, T. Palaniyandi, R. Shanmugam, S. Karunakaran, M. Pandi, M.R.A. Wahab, G. Baskar, B.K. Rajendran, A. Sivaji, M. Moovendhan, Synthesis, characterization, cytotoxicity, and antimicrobial studies of green synthesized silver nanoparticles using red seaweed *Champia parvula*. Biomass Convers. Biorefin. (2023)

S. Viswanathan, T. Palaniyandi, R. Shanmugam, M. Tharani, B.K. Rajendran, A. Sivaji, Biomedical potential of silver nanoparticles capped with active ingredients of *Hypnea valentiae*, red algae species. Part. Sci. Technol. **40**, 686–696 (2022)

M. Vivek, P.S. Kumar, S. Steffi, S. Sudha, Biogenic silver nanoparticles by *Gelidiella acerosa* extract and their antifungal effects. Avicenna J. Med. Biotechnol. **3**, 143–148 (2011)

N. Willian, S. Syukri, Z. Zulhadjri, H. Pardi, S. Arief, Marine plant mediated green synthesis of silver nanoparticles using mangrove *Rhizophora stylosa*: effect of variable process and their antibacterial activity. F1000Res **10**, 768 (2022)

N. Willian, Z. Syukri, A. Labanni, S. Arief, Bio-Friendly synthesis of silver nanoparticles using mangrove *Rhizophora stylosa* leaf aqueous extract and its antibacterial and antioxidant activity. Rasayan J. Chem. **13**, 1478–1485 (2020)

References

Y.P. Yew, K. Shameli, M. Miyake, N.B.B.A. Khairudin, S.E.B. Mohamad, T. Naiki, K.X. Lee, Green biosynthesis of superparamagnetic magnetite Fe_3O_4 nanoparticles and biomedical applications in targeted anticancer drug delivery system: a review. Arab. J. Chem. **13**, 2287–2308 (2020)

Y.P. Yew, K. Shameli, M. Miyake, N. Kuwano, N.B.B.A. Khairudin, S.E.B. Mohamad, K.X. Lee, Green synthesis of magnetite (Fe_3O_4) nanoparticles using seaweed open access (*Kappaphycus alvarezii*) extract. Nanoscale Res. Lett. **11**, 276 (2016)

M. Yousefzadi, Z. Rahimi, V. Ghafori, The green synthesis, characterization and antimicrobial activities of silver nanoparticles synthesized from green alga *Enteromorpha flexuosa* (wulfen). J. Agardh. Mater Lett **137**, 1–4 (2014)

Y.A. Yugay, R.V. Usoltseva, V.E. Silant'ev, A.E. Egorova, A.A. Karabtsov, V.V. Kumeiko, S.P. Ermakova, V.P. Bulgakov, Y.N. Shkryl, Synthesis of bioactive silver nanoparticles using alginate, fucoidan and T laminaran from brown algae as a reducing and stabilizing agent. Carbohydr. Polym. **245**, 116547 (2020)

M.K. Zahran, H.A. Mohammed, Green synthesis of silver nanoparticles using polysaccharide extracted from *Laurencia obtusa* algae. Egypt. J. Appl. Sci. **36**, 9–16 (2021)

D. Zhang, X. Ma, Y. Gu, H. Huang, G. Zhang, Green synthesis of metallic nanoparticles and their potential applications to treat cancer. Front. Chem. **8**, 799 (2020a)

D. Zhang, G. Ramachandran, R.A. Mothana, N.A. Siddiqui, R. Ullah, O.M. Almarfadi, G. Rajivgandhi, N. Manoharan, Biosynthesized silver nanoparticles using *Caulerpa taxifolia* against A549 lung cancer cell line through cytotoxicity effect/morphological damage. Saudi J. Biol. Sci. **27**, 3421–3427 (2020b)

H. Zhang, J.A. Jacob, Z. Jiang, S. Xu, K. Sun, Z. Zhong, N. Varadharaju, A. Shanmugam, Hepatoprotective effect of silver nanoparticles synthesized using aqueous leaf extract of *Rhizophora apiculata*. Int. J. Nanomedicine **14**, 3517–3524 (2019)

The manufacturer's authorised representative in the EU is Springer Nature Customer Service Centre GmbH, Europaplatz 3, 69115 Heidelberg, Germany. If you have any concerns regarding our products, please contact ProductSafety@springernature.com

Printed and bound by CPI Group (UK) Ltd, Croydon, CR0 4YY

26/03/2026

02078983-0005